T0074608

3D Image Reconstruction for CT and PET

CRC Press Focus Series in Medical Physics and Biomedical Engineering

Series Editors:
Magdalena Stoeva and Tae-Suk Suh

Recent books in the series:

3D Image Reconstruction for CT and PET: A Practical Guide with Python
Daniele Panetta and Niccolò Camarlinghi

Computational Topology for Biomedical Image and Data Analysis
Rodrigo Rojas Moraleda, Nektarios Valous, Wei Xiong, Niels Halama

e-Learning in Medical Physics and Engineering: Building Educational Modules with Moodle
Vassilka Tabakova

3D Image Reconstruction for CT and PET

A Practical Guide with Python

Dr. Daniele Panetta

Dr. Niccolò Camarlinghi

CRC Press

Taylor & Francis Group

Boca Raton London New York

CRC Press is an imprint of the
Taylor & Francis Group, an **informa** business

First edition published 2020
by CRC Press
6000 Broken Sound Parkway NW, Suite 300, Boca Raton, FL 33487-2742

and by CRC Press
2 Park Square, Milton Park, Abingdon, Oxon, OX14 4RN

© 2020 Taylor & Francis Group, LLC

CRC Press is an imprint of Taylor & Francis Group, LLC

Library of Congress Cataloging-in-Publication Data
Names: Panetta, Daniele, author.
Title: 3D image reconstruction for CT and PET : a practical guide with
    Python / Daniele Panetta, Camarlinghi Niccolo.
Description: Boca Raton : CRC Press, 2020. | Series: Focus series in
    medical physics and biomedical engineering | Includes bibliographical
    references and index.
Identifiers: LCCN 2020017206 | ISBN 9780367219970 (hardback) | ISBN
    9780429270239 (ebook)
Subjects: LCSH: Tomography--Data processing. | Tomography, Emission--Data
    processing. | Three-dimensional imaging. | R (Computer program language)
Classification: LCC RC78.7.T6 P36 2020 | DDC 616.07/57--dc23
LC record available at https://lccn.loc.gov/2020017206

ISBN: 9780367219970 (hbk)
ISBN: 9780429270239 (ebk)

Typeset in LMRoman
by Nova Techset Private Limited, Bengaluru & Chennai, India

To Alessandro,
your curiosity and your smile push me
every day to be better.

To Davide,
my little big champion, who is three years old,
and pretends to be a doctor.

To Giulia,
for the joy she gives me, every time I open the
door, when she runs to me with big smiles.

This book is dedicated to our beloved children, Alessandro Camarlinghi, Davide Panetta, and Giulia Panetta.

Contents

Preface

WHAT DOES THIS BOOK DEAL WITH?

This book focuses on image reconstruction from projections, with particular emphasis on Computed Tomography (CT) and Positron Emission Tomography (PET). Even though this topic is often seen as a branch of pure mathematics, referred to as *Tomographic Reconstruction*, it is indeed a strongly multidisciplinary domain involving, physics, engineering, computer science and, of course, any discipline relevant for the final application (not just medical) for which the above mentioned imaging modalities are used. This multidisciplinary view of image reconstruction is even more evident when working in research contexts, like the authors of this book do. In this context, especially when developing new technology, with custom geometry and diverse (sometimes highly heterogeneous) hardware configurations, it is very important to master not just the theory behind image reconstruction, but also its practical aspects with custom software.

WHO IS THIS BOOK FOR?

This book is meant for any student, researcher, or engineer who wants to start taking first steps in the field of image reconstruction for CT and PET. Most of the content is based on the authors' lectures, delivered to the students of the Medical Physics program at the University of Pisa, Italy. The level of prior knowledge required for reading this book is that generally required during the last year of an undergraduate program in physics, mathematics or engineering. A background of scientific programming in Python/Numpy is also necessary to get fruitful understanding of the sample code, listings and Jupyter notebooks provided as companions, even though this is not required for the comprehension of the theory itself. This book can also be useful for readers who are already familiar with the theory of image reconstruction from projections, but are still in search of a beginner-level set of (editable) software tools to start their journey into the practice.

WHAT THIS BOOK IS NOT

This book is neither a tutorial, nor a documentation of a specific software library for image reconstruction. There are several well-implemented, nicely documented, and computationally efficient software libraries available for this purpose. Just to mention a few of them:

- ASTRA toolbox (MATLAB and Python, open source, www.astra-toolbox.com/)
- STIR (C++, open source, stir.sourceforge.net/homepage.shtml)
- RTK (C++, open source, www.openrtk.org/RTK/resources/soft-ware.html)
- TIGRE (MATLAB and Python, open source, github.com/CERN-/TIGRE)
- COBRA (Exxim CC, USA, C++, commercial, www.exxim-cc.com/products_cobra.htm)
- CASToR (C++, open source, www.castor-project.org/)

(this list may be incomplete). The purpose of the modules, scripts and notebooks provided along with this book is definitely not to compete with the libraries listed above. As already mentioned, the goal of this book is to create a bridge between theory and practice; we strongly believe that, in the specific field addressed by this book, having a set of easy-to-read (and easy-to-run, but not necessarily performant) software samples is the key for the beginner to start fruitful practice in image reconstruction. Due to the complexity of other industry-level programming languages (e.g., C/C++), but also because of current trends in scientific programming, the choice of Python/Jupyter was quite natural for the purpose of this book.

HOW DOES THIS BOOK COMPARE WITH OTHERS IN THE SAME FIELD?

There are of course several comprehensive textbooks on tomographic reconstruction, and we will often refer to them in the subsequent chapters. In this book, we have decided to put the reader much closer to the practical side of image reconstruction, rather than on the math/ theory side. That is, we tried (and let's hope we succeeded!) to create a bridge between the rigor of the mathematical theory of image reconstruction, described in depth in the texts and articles that we will often refer to, and the practical application with simple ready-to-go software samples. Such samples can be tested, edited and modified by the reader in order to enable him or her to gain immediate understanding of how the relevant parameter affects the final image. The choice of the programming language (Python) was mainly dictated by a mix of ease of implementation, availability of high-level numerical libraries (such as *Numpy/Scipy*), platform-independence, and last but not least, popularity among the community of scientific programmers. The implementation of the Python scripts is not optimized for performance (they might run quite slowly as compared with other libraries), but performance was not the goal here. The focus was, of course, on code readability and ease of comprehension. We hope that the readers will appreciate the effort that we have invested to make this bridge as smooth as possible.

STRUCTURE OF THE BOOK

The book is structured as follows: first, a concise overview of the main concepts, definitions and notions required in the description of the algorithms will be given in Chapter 1. Chapter 2 describes the structure of the companion Python library (DAPHNE) and its usage in the examples provided along with the book. Practical information on how to install and run this software is also provided in this chapter. The subsequent chapters will deal with the selected image reconstruction algorithms. Chapter 3 will focus on Fourier-based methods, mainly FBP, in the most common geometries relevant for CT and PET (2D parallel beam) or just CT (2D fan beam and 3D cone beam). Chapter 4 deals with iterative methods (ART, SIRT, MLEM and OSEM). A last chapter is dedicated to the topic of synthetic generation of projection data.

About the Authors

Daniele Panetta is a researcher at the Institute of Clinical Physiology of the Italian National Research Council (CNR-IFC) in Pisa. He earned his MSc degree in Physics in 2004 and his specialisation diploma in Health Physics in 2008, both at the University of Pisa. From 2005 to 2007, he worked at the Department of Physics "E. Fermi" of the University of Pisa in the field of tomographic image reconstruction for small animal imaging micro-CT instrumentation. His current research at CNR-IFC has as its goal the identification of novel PET/CT imaging biomarkers for cardiovascular and metabolic diseases. In the field of micro-CT imaging, his interests cover applications of three-dimensional morphometry of biosamples and scaffolds for regenerative medicine. He acts as reviewer for scientific journals in the field of medical imaging: Physics in Medicine and Biology, Medical Physics, Physica Medica, and others. Since 2012, he has been adjunct professor in Medical Physics at the University of Pisa.

Niccolò Camarlinghi is a researcher at the University of Pisa. He obtained his MSc in Physics in 2007 and his PhD in Applied Physics in 2012. He has been working in the field of Medical Physics since 2008 and his main research fields are medical image and analysis and image reconstruction. He is involved in the development of clinical, pre-clinical PET and hadron therapy monitoring scanners. He is currently a lecturer at University of Pisa, teaching courses of life-science and medical physics laboratory. He regularly acts as a referee for the following journals: Medical Physics, Physics in Medicine and Biology, Transactions on Medical Imaging, Computers in Biology and Medicine, Physica Medica, EURASIP Journal on Image and Video Processing, Journal of Biomedical and Health Informatics.

## CHAPTER 1

Preliminary notions

The purpose of this chapter is to provide a concise list of the most important theoretical and practical elements involved in CT and PET reconstruction, along with a brief description for each of them. The list does not follow a precise logical order, and it is constructed with the ultimate goal to make the reading of the following chapters as fluent and goal-oriented as possible (with the "goal" being the reconstruction of tomographic images).

Those readers who are already familiar with the definitions and concepts provided in this chapter, can skip reading it. For those who are still at the preliminary stage of their study in PET/CT technology and reconstruction, the reading of this chapter is mandatory for a full understanding of what comes next. Moreover, it will serve as a quick reference during reading of the subsequent chapters, for all those concepts that are referenced therein. The beginners can then come back to this chapter at any time to refresh their knowledge.

1.1 IMAGE RECONSTRUCTION FROM PROJECTION

1.1.1 Purpose of image reconstruction

Given an unknown function f describing the spatio-temporal distribution of some chemical-physical property of a patient or object under examination, and a set p of projections of f (the formal definition of projection will be given below when discussing the concepts of line integral and ray-sum), the purpose of image reconstruction is to recover the best estimate of f from p. This problem is referred to as an inverse problem, in which the final goal is to retrieve an unknown input data back from the result (output) of a physical measurement performed on it. In the domain of tomographic imaging, the corresponding forward problem is precisely the process of acquiring the projection data, by means of specific instrumentation for which technical description is well beyond of the scope of this book. Even though any forward problem has a defined solution, this is not true for inverse problems. That is, several possible images are compatible with a given set of projection measurements. Hence, image reconstruction is referred to as an ill-posed inverse problem. For a thorough discussion on inverse problems in imaging, refer to Bertero and Boccaccio [2].

1.1.2 Families of reconstruction methods

Reconstruction methods in tomographic imaging can be grouped into two families:

- Analytical reconstruction methods

- Iterative reconstruction methods

Analytical methods treat both images and projections as continuous functions. The process of image acquisition is ideally represented as a linear and continuous operator acting on the space of object functions representing the patient (or object) under examination. Hence, the acquisition process gives rise to a integral transform of the unknown function describing the patient; image reconstruction must be carried out by inverting such an integral transform. The discrete nature of projection data and reconstructed image (in digital form) is taken into account only at the very final stage of image reconstruction, when a closed analytical inversion formula has been derived starting from the continuous assumption. Iterative algorithms are instead based on the assumption that both images and projection data are discrete. In this view, the forward problem of projection data acquisition can be conveniently set in the form of a system of linear equations. Solving such a system of equations is not a trivial task in real-life situations, given that both the number of unknowns (the voxel values) and the number of equations (their projections) can easily go beyond several millions or billions of elements. Hence, a generalized inversion of the system of equations is sought, by means of iterative techniques. Even though these methods are more computationally intensive as compared to analytical ones, they can lead to improved performance over analytic methods, as they offer the chance to model accurately the physics of the signal generation and detection [30, 13]. Moreover, they can be easily adapted to different geometries without major modifications. We can further divide iterative algorithms into two sub-categories: algebraic and statistical. Algebraic algorithms model the reconstruction problem as a noiseless linear system, thus neglecting the intrinsic stochastic nature of radiation generation and detection. This is usually not a limit when statistical fluctuations are negligible with respect to the detected signal, e.g. for regular dose CT. In the context of PET or low dose CT the relative fluctuation can be relevant. In other words, this means that you will never acquire the same data twice even if the experimental setup is exactly the same! In this situation, modeling the statistical nature of the acquired data can improve the reconstructed image quality. Statistical algorithms are built on hypothesis on the data probabilistic distribution to build a likelihood function whose maximization will provide the solution to the image reconstruction problem. Both algebraic and statistical methods suffer from a higher computational demand with respect to analytical methods. This is due to the need to perform several iterations, each involving forward and backprojection operations. This forced the researchers into a space-time trade-off that limited the application of these methods before the last two decades.

1.2 TOMOGRAPHIC IMAGING MODALITIES (RELEVANT FOR THIS BOOK)

1.2.1 Computed Tomography (CT)

CT is a tomographic imaging technique based on X-ray transmission through the object under examination. In the most recent clinical implementations, it uses a divergent beam of radiation emitted from an X-ray tube and a curved 2D array solid-state detectors arranged in pixels. Both the source and the detectors are placed on a rotating gantry, allowing the acquisition of projection data of the object under examination from any angle.

CT configurations based on flat-panel detectors are also possible. This geometry is mostly used for dental imaging, intraoperative imaging, image guidance of radiotherapy treatments, as well as in small animal (preclinical) imaging, industrial imaging and for benchtop micro-CT's (or μ-CT) in which the object is rotating around a vertical axis, whereas the source-detector system is still in the laboratory frame of reference. Object rotation is also employed for advanced imaging at synchrotron facilities, where the useful X-ray beam cannot be re-oriented.

1.2.2 Positron Emission Tomography (PET)

Positron-emission tomography (PET) is a functional imaging technique that can be used to observe metabolic processes in a living organism. PET allows the study of the spatio-temporal distribution of special radio-labelled molecules that are injected in a living organism. A radio-labelled molecule is obtained by substituting or adding a β^+ unstable nucleus with a relatively short half life (typically in the order of 10-100 min for the most used nuclides). Once the radioactive nucleus decays it emits a positron, which loses its kinetic energy in the medium and eventually annihilated with an electron. The annihilation of the positron with an atomic electron in the patient gives rise to the emission of a pair of γ photons, approximately 180 degrees apart, each with an energy of 511 keV. The line where the two photons travel is often referred to as line of flight (LOF) (see Figure 1.1). Each pair of annihilation photons is detected by a system of position sensitive γ detectors, typically arranged in a cylindrical configuration around the object under examination. The detection of a given photon pair is made possible by a system of electronic coincidence with timing windows in the order of a few nanoseconds. This is sometimes called "electronic collimation" as opposed to physical collimation of SPECT. Since emission is isotropic, by detecting the photon pairs we can reconstruct the spatio-temporal distribution of the tracer that was injected.

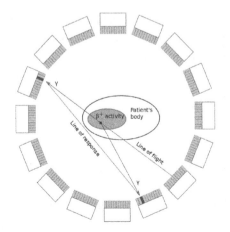

Figure 1.1: Picture representing the difference between LOF and LOR in a typical PET detector.

Due to the statistical nature of interaction of photons with matter, the actual LOF cannot be reconstructed with certainty. The reconstructed line is called line of response (LOR).

Data acquired with PET can either be arranged as list-mode, i.e. a list of all the detected coincidences, sinogram (see 1.4.3) or histogram, i.e. each LOR appears together with its multiplicity. In this book we will assume that LOR data are stored in histogram mode. Non-cylindrical configuration of PET detectors is also possible, for instance in special applications such as hadron therapy range monitoring (where planar detection heads are used) or small animal PET (typically employing detector rings made of few edges). For a more comprehensive description of positron emission tomography the reader can refer to [4, 33].

1.2.3 Single-photon Emission Computed Tomography (SPECT)

SPECT is the tomographic imaging modality strictly linked to the planar *scintigraphy*, a very well known nuclear medicine modality. We could say that SPECT is the 3D counterpart of scintigraphy, somewhat similarly to how CT is the 3D counterpart of radiography. Unlike PET, the detection of gamma photons in SPECT relies on the usage of dedicated collimators, which have the disadvantage of decreasing the number of detected photons and hence the sensitivity of the methods. The first generation of SPECT scanners used to be based on the so-called Anger camera configuration [1, 25], employing layers of inorganic scintillators (typically NaI) coupled to an array of photomultiplier tubes (PMT). More recent configurations are instead based on solid state detectors (mostly cadmium-zinc telluride, or CZT), allowing more flexible geometrical designs as well as increased energy resolutions.

Depending on the shape of the collimator holes, the acquisition geometry of a SPECT system can resemble those of first generations CT systems (i.e., parallel beam, when parallel hole collimators are employed), or also cone beam geometries when pinholes are used. Hence, the problem of 2D/3D reconstruction in SPECT is in most cases similar to CT; unlike CT, in SPECT there is the peculiar need to address the problem of gamma ray attenuation into the patient's body. This topic will not be covered in this book: this is part of the reason why we didn't add the name "SPECT" in the book title. Nevertheless, most of the results and algorithms discussed in this book apply very well also to SPECT data, provided that a suitable preprocessing for photon attenuation has been applied prior to the reconstruction step.

1.3 NOTIONS COMMON FOR ALL RECONSTRUCTION METHODS

1.3.1 Object function and image function

From a mathematical point of view, any imaging device can be modeled ideally as a linear operator S taking a function $f(x, y, z, t)$ as its input (the object) and returning a second function $g(x, y, z, t)$ as output (the image):

$$g = Sf. \tag{1.1}$$

The imaging system S:

- is *linear* if, given an object function of the form $f = \sum a_i f_i$, the output is of the form $g = S \sum a_i f_i = \sum a_i S f_i$;

- is *stationary* if its response does not change over time;

- is *shift-invariant* if its response does not change across the support of the object function f.

Most real-world PET and CT scanners are shift-variant (the response changes from the center to the periphery of the field of view), are with good approximation stationary.

It is important to stress that the object function f only depends on the chemical-physical characteristics of the object (or patient) under examination, and it's not related to the performance of the imaging tool used to image it. In CT, this function ultimately represents the spatio-temporal distribution the patient's attenuation coefficients (usually denoted as μ), whereas in PET the value of f represents a local concentration of activity inside the body. The image function g can be thought of as the way the imaging system S is "seeing" f, which of course includes blurring, noise and artifacts depending on hardware performance, acquisition protocol and reconstruction method and parameters.

1.4 RELEVANT NOTIONS FOR ANALYTICAL RECONSTRUCTION METHODS

1.4.1 Line integral

The integral of the object function f along a given ray path is referred to as the line integral. The concept of line integral is strictly linked to the available data in a reconstruction task. In CT, the line integral represents the total attenuation of the X-rays along a beam direction connecting the X-ray source (modeled as a geometric point) and a given pixel of the detector array. In PET, the concept of line integral can be used as well as it represents the integrated activity concentration of the object along a given LOR (see Section 1.2.2).

1.4.2 Radon transform

In two dimensions, the Radon Transform (RT) of a function f is another function $p = \mathcal{R}f$ representing a complete set of line integrals of f, for all possible lines intersecting the object. Given a line of equation

$$x \cos \phi + y \sin \phi - x' = 0, \tag{1.2}$$

the RT of a function f in the xy plane can be written as

$$p(x', \phi) = \int\limits_{-\infty}^{\infty} \int\limits_{-\infty}^{\infty} f(x, y) \delta_1(x \cos \phi + y \sin \phi - x') \, \mathrm{d}x \mathrm{d}y, \tag{1.3}$$

where δ_1 is the 1-dimensional Dirac delta function. In the n-dimensional case, with $n > 2$, the RT is no longer the set of line integrals of f but instead a set of plane integrals, through hyperplanes with dimension $n - 1$. Among its various properties, the RT is linear and it represents in a idealized fashion the process of projection data acquisition in real imaging systems.

1.4.3 Sinogram

The graphical representation of the RT of an object function f (Section 1.4.2) in 2D is often referred to as the sinogram of f. The reason for this name is that the 2D RT maps each point (x, y) of the spatial domain into a portion of space of the Radon domain (x', ϕ) having a sinusoidal shape. Due to the linearity of the RT, the sinogram of any object f is the superposition of several sinusoids (each with different phase, amplitude and intensity) related to each point of the object function. Hence, sinograms are (generally) not understandable by human observers; in the context of analytical methods, the final goal of image reconstruction is to recover images from their sinograms.

1.4.4 Exact and approximated reconstruction

An analytical reconstruction algorithm is referred to as *exact* when, in absence of noise, the error between the reconstructed image and the original object can be made arbitrarily small just by refining the spatial (radial and angular) sampling in the acquisition process. In all other cases, the algorithm is referred to as *non-exact* (or approximated).

1.4.5 Central section theorem

Let us consider the two coordinate systems Ouv and $Ou'v'$ in the frequency domain, associated with the laboratory and gantry, respectively. For a projection p acquired at angle ϕ, we denote by $P = \mathcal{F}_1 p$ its 1D Fourier transform (FT) with respect to the radial coordinate x':

$$
P(\nu, \phi) = \int_{-\infty}^{\infty} p(x', \phi) \, e^{-j2\pi\nu x'} \, dx'
$$

$$
= \int_{-\infty}^{\infty} \int_{-\infty}^{\infty} f(x', y') \, e^{-j2\pi\nu x'} \, dx' dy' , \qquad (1.4)
$$

where ν is the frequency variable associated with x'. In this context, it is useful to express the polar coordinates of the frequency domain in a signed form, i.e., $\nu \in \,]-\infty, \infty[$ and $\phi \in [0, \pi[$.

We can now look for some relationship between the above 1D FT of the projection and the 2D FT of the object function. To this purpose we rewrite the function $F = \mathcal{F}_2 f$ in the gantry coordinate system as

$$
F(\xi', v') = \int_{-\infty}^{\infty} \int_{-\infty}^{\infty} f(x', y') \, e^{-j2\pi(\xi' x' + v' y')} \, dx' dy' . \qquad (1.5)
$$

By comparing Equations 1.5 and 1.4 and arguing that in signed polar coordinates $\xi' = \nu$, it is easy to show that

$$
F(\xi', 0) = P(\nu, \phi) \qquad (1.6)
$$

or equivalently, by using the laboratory coordinates,

$$
F(\xi, v)\big\rvert_{v \cos\phi - \xi \sin\phi = 0} = P(\nu, \phi) . \qquad (1.7)
$$

The restriction of the function F to the line of equation $-\xi \sin \phi + v \cos \phi = 0$ (or, simply, $v' = 0$) is the central section of F, taken at angle ϕ. Equations 1.6 and 1.7 are two equivalent statements of the central section theorem (CST). In the literature, this theorem is also referred to as the Fourier slice theorem.

The theorem just stated is a cornerstone of analytical image reconstruction as it provides a direct link between the object and its projections in the frequency domain. Reconstruction algorithms based on the CST fall in the category of the so-called Fourier-based methods. As a practical interpretation of CST, we could say that due to Equation 1.7 each projection of f at an angle ϕ gives us access to a "sample" (i.e., a central section) of the 2D FT of f. Intuitively, by taking as many independent projections as possible, we can progressively refine our knowledge of F, and eventually reconstruct f by taking the inverse 2D FT of such approximated (due to the finite sampling) estimate of F. We will see in the following section that the approach just mentioned leads to one possible solution to the problem of image reconstruction, called Direct Fourier Reconstruction (DFR). Figure 1.2 shows the link between p and f, in both the space and frequency domains.

1.5 RELEVANT NOTIONS FOR ITERATIVE RECONSTRUCTION METHODS

1.5.1 Object vector and data vector

Iterative algorithms employ a discrete representation of both projection data and image data. In this book we will always assume that the image and the projection are respectively discretised into N and M elements. The most common way to represent the object function is

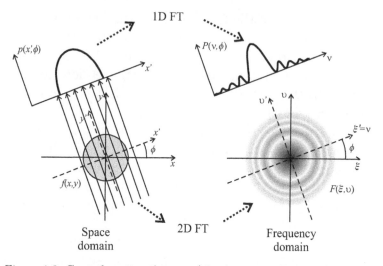

Figure 1.2: Central section theorem (also known as Fourier slice theorem). The 1D FT of a row of the parallel beam sinogram of f taken at angle ϕ is equal to a central section of the 2D FT of F taken at the same angle.

in the form of a vector of non-overlapping cubic basis functions, i.e., voxels

$$\mathbf{f} = (f_1, f_1, ..., f_N) \tag{1.8}$$

whereas the projection data can be discretised into a vector of M components

$$\mathbf{p} = (p_1, p_1, ..., p_M). \tag{1.9}$$

We will also define the Target Region (TR) as the physical region that we aim to reconstruct.

1.5.2 System matrix

The system matrix (SM) is the mathematical entity connecting the discrete representation of the object function to the discrete representation of the projection data, i.e.,

$$\mathbf{p} = \mathbf{A} \cdot \mathbf{f} \tag{1.10}$$

where $\mathbf{A} \in \mathbb{R}^{M \times N}$ is the SM. Alternatively we can write Equation 1.10 as

$$
\begin{vmatrix} p_1 \\ p_2 \\ .. \\ p_M \end{vmatrix} = \begin{vmatrix} A_{1,1} & A_{1,2} & .. & A_{1,N} \\ A_{2,1} & A_{2,2} & .. & A_{2,N} \\ .. & .. & .. & .. \\ A_{M,1} & A_{M,2} & .. & A_{M,N} \end{vmatrix} \cdot \begin{vmatrix} f_1 \\ f_2 \\ .. \\ f_N \end{vmatrix}
\tag{1.11}
$$

where the element $A_{i,j}$ is the probability that the j-th voxel contributes to the i-th projection. Therefore, the SM can, in principle, account for the all the physics ruling the projection generation. The k-th row of $\mathbf{A}$ (denoted as $\mathbf{A}_k$) is often called the k-th "tube of response" (TOR) of the SM. This terminology is due to the fact that the only non-null elements are those within a tube connecting the two extrema of the k-projection. Direct inversion of Equation 1.10 is not possible/useful for two reasons:

- The SM matrix is not square in general.

- Pseudo inverse of the SM has a large conditioning number, i.e. a small perturbation in $\mathbf{p}$ will generate a large variation in the reconstructed image $\mathbf{f}$.

Moreover, due to its size the SM cannot be stored as a dense matrix in typical cases: for example, let's consider an imaging system where $N \sim 10^6$ and $M \sim 10^7$ (such numbers are not very large nowadays). The SM of such an imaging system, would contain 10^{13} elements and it would take, at least, 40 terabytes to store them all as four bytes floating point. For this reason handling the SM is not trivial. In order to be able to use the SM we can either store it on file or compute the part of the matrix we need "on the fly". Storing the SM on file requires reducing its footprint by exploiting its sparseness, i.e. storing only non-null elements or reducing its redundancy. These aspects are outside the scope of this book and will not be discussed. As discussed in [13] a realistic SM can improve image quality. However, we would like to stress that the following quote applies nicely to this context: "all the models are wrong but some are useful" (George E. P. Box). This means that no SM can account entirely for all the physical effects present in a imaging system; nonetheless those SM that model the most relevant effects in the context are expected to produce better results.

1.5.3 Discrete forward projection

In the context of iterative reconstruction, the line integral becomes a discrete summation. The k-th line integral (q_k) can be expressed as

$$q_k[\mathbf{f}] = \sum_{i=1}^{N} A_{k,i} f_i. \tag{1.12}$$

A whole set of line integrals can be expressed as a function of the SM as

$$\mathbf{q}[\mathbf{f}] = \mathbf{A} \cdot \mathbf{f} \tag{1.13}$$

where $\mathbf{q}$ is a vector containing all the line integrals. We will call the vector $\mathbf{q} \in \mathbb{R}^M$ the "forward projection of $\mathbf{f}$". The forward projection is therefore an operation $\mathbb{R}^N \to \mathbb{R}^M$, i.e., it goes from image space to projection space. We can think of the forward projection as an operator that gives us an estimation of the projections generated by the object function $\mathbf{f}$.

1.5.4 Discrete back projection

The back projection of $\mathbf{v} \in \mathbb{R}^M$ is defined as

$$\mathbf{b}[\mathbf{v}] = \mathbf{A}^{\mathbf{t}} \cdot \mathbf{v} \tag{1.14}$$

$\mathbf{A}^{\mathbf{t}}$ denotes the transpose of the SM. The back projection is therefore an operation $\mathbb{R}^M \to \mathbb{R}^N$, i.e., it goes from projection space to image space. We will also use the operator $\mathbf{b}_k[z]$, with $z \in \mathbb{R}$, to denote the back projection of the k-th TOR as

$$\mathbf{b}_k[z] = A_k^t \cdot z. \qquad (1.15)$$

Moreover, Equation 1.14 can be expressed as

$$\mathbf{b}[\mathbf{v}] = \sum_{i=1}^{M} A_i^t \cdot v_i \qquad (1.16)$$

which means that the back projection of $\mathbf{v}$ can be expressed as a linear combination of the TOR of the SM, i.e., the A_i. This shows the feasibility of performing a back projection operation even when computing the rows of the SM one at time.

## CHAPTER 2

Short guide to Python samples

2.1 INSTALLATION

This book comes with a software framework written in Python language. We called it DAPHNE (Didactic tomogrAPHic recoNstruction framEwork), with the specific goal to let the user perform simple experiments with the most common reconstruction algorithms for PET and CT. In order to be able to use DAPHNE, the reader must follow the instruction contained in the Readme.md file (downlodable from the repository located at www.routledge.com/9780367219970).

2.2 PROJECT ORGANIZATION

DAPHNE is organized in three packages:

- Algorithms

- Geometry

- Misc

The reconstruction algorithms and other helper classes (such as SiddonProjector and ProjectionData-Generator) are implemented in the Algorithms package.

The **Geometry** package contains all the functions required to generate and handle CT and PET experimental setups. In the **Misc** module, we have placed all the utilities that do not fit in the other two packages. The Data folder contains example images and projection data, useful in performing some of the suggested experiments. However, it is desirable that the readers try to perform their own experiments with other data. Finally, in the **Notebook** folder we have put a collection of Jupyter notebooks [17] that demonstrate how to run some of the examples shown hereafter.

2.3 CODING CONVENTIONS

The code contained in this project implements simple conventions, with the main intent to improve its readability:

- The names of class methods are camelcase.

- The names of class attributes uses underscore (_) as a word separator.

- Class attributes beginning with _, are intended to be private, i.e., should not be modified or accessed directly by the user.

- The set/get class methods are not implemented unless they are non-trivial.

- The name of the attributes contain as last token the unit of measure of the variable, i.e., _mm: millimeters, _deg: degrees, _nb: pure number.

2.4 DEFINITION OF AN EXPERIMENTAL SETUP

In order to be able to reconstruct images, the user must define the image modality used, i.e., CT or PET, the Target Region (TR), i.e., the portion of space to be reconstructed, and the position of detector pixels and radiation sources. The experimental setup is fully described by the two classes `ExperimentalSetupCT` and `ExperimentalSetupPET`. Both classes are derived from the generic `ExperimentalSetup` class, and implement experimental setups for a CT and a PET imaging system, respectively. The main attributes defined in an object of type `ExperimentalSetupCT` or `ExperimentalSetupPET` are:

- The imaging modality, i.e. CT or PET.

- In case of CT, the detector shape (planar or arc).

- The physical positions of the detector pixels (`_pixel_pos_mm`).

- The physical positions of the radiation sources (`_sources_pos_mm`).

- The physical size of the TR (`image_matrix_size_mm`) and its sampling step (`voxel_size_mm`).

- The physical positions used to cast projections (`_projection_extrema`).

2.4.1 Definition of a radiation detector

The set of all the pixels contained in a experimental setup represents the radiation detector. Instead of placing manually each source and pixel in the 3D space, the user can set the following parameters into an experimental setup instance:

- `detector_pixels_per_slice_nb`: number of pixels per detector row.

- `detector_slices_nb`: the number of detector slices or rows.

- `detector_slice_pitch_mm`: the distance Δz in mm between the center of two contiguous slices or rows.

2.4.2 Definition of the image matrix

The region of space to be reconstructed, i.e., the TR, and its sampling, can be defined by setting the following variable in a instance of the ExperimentalSetup:

- image_matrix_size_mm: 3 element array containing the size of TR in mm along each direction. As the TR center is always assumed to be in the point (0,0,0), it spans [−image_matrix_size_mm/2,image_matrix_size_mm/2] in each direction.

- voxel_size_mm: 3 element array containing the size of a voxel in mm along each direction.

The number of voxels per direction of the image matrix is calculated as:

_voxels_nb = image_matrix_size_mm//voxel_size_mm.

The left hand coordinate system shown in Figure 2.1 is used. The origin of the axis is always coincident with center of the reconstructed zone. Once the parameters of an experimental setup are set, the user must call the Update

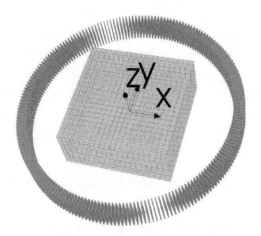

Figure 2.1: Rendering of a PET experimental setup: pixels are shown in dark gray, whereas the TR is shown in light gray.

method. This triggers the computation of all the private variables described in Section 2.4.

2.4.3 PET experimental setup

The PET experimental setup defines a cylindrical detector (see Figure 2.1), made of pixels stacked on different slices. To define how the _projection_extrema are computed, the h_fan_size is used: each crystal is set to acquire coincidences with the 2*h_fan_size+1* detector_slices_nb in front of it. Since PET is an emission technique, no source has to be defined.

2.4.4 CT experimental setup

To define an experimental setup containing a CT scanner, the user must choose between three different irradiation modalities:

- parallel beam irradiation

- fan beam irradiation

- cone beam irradiation.

Moreover, when using fan and cone beam modes, it is possible to choose between two different detector shapes: planar and arc. In every CT experimental setup, the detector is assumed to rotate around the (0,0,0) point, on a axis parallel to z. The parameters angular_range_deg and gantry_angles_nb are used to set the rotation range and the number of rotation steps that are used to generate the projections.

The parameters to be defined for a CT detector depend on the mode used and are summarised in Table 2.1.

2.4.5 Parallel beam CT

In this geometry, the number of sources is equal to the number of detector pixels. Each source irradiates only

Table 2.1: Required attributes for the different CT geometries.

param name	parallel beam	fan beam	cone beam
pixels_per_slice_nb	✓	✓	✓
detector_slices_nb	✓	✓	✓
slice_pitch_mm	✓	✓	✓
angular_range_deg	✓	✓	✓
gantry_angles_nb	✓	✓	✓
fan_angle_deg	✗	✓	✓
sad_mm	✗	✓	✓
sdd_mm	✗	✓	✓

the pixel in front of it. Due to the nature of this geometry, the distance between a source and its target pixel is not relevant. Figure 2.2(a) shows a rendering of a parallel beam CT detector when the gantry angle is set to 0.

2.4.6 Fan beam CT

In this geometry the number of sources is equal to the number slice of the detector. Figure 2.2(b) shows a rendering of a fan beam CT detector when the gantry angle is set 0. Each source irradiates a fan of angular size fan_angle_deg. The two main parameters that define this geometry are:

- sdd_mm: Source to Detector Distance in mm (SDD).

- sad_mm: Source to Axis Distance in mm (SAD). With sdd_mm > sad_mm.

The size of the detector is evaluated by computing $2 \cdot \tan(\mathtt{fan\_angle\_deg}/2) \cdot \mathtt{sdd\_mm}$.

2.4.7 Cone beam CT

In this geometry only one source needs to be placed. The source irradiates a cone in front of it. Figure 2.2(c) shows a rendering of a cone beam CT detector when the

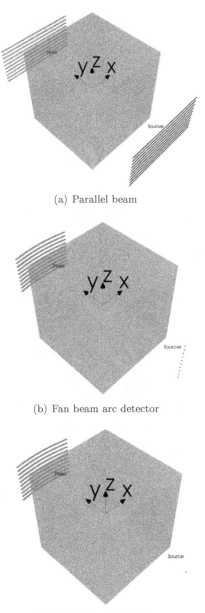

(a) Parallel beam

(b) Fan beam arc detector

(c) Cone beam arc detector

Figure 2.2: Rendering of parallel, fan, cone beam setups when the gantry angle is equal to 0. Pixels and sources are shown in dark gray and the TR is shown in light gray.

gantry angle is 0. The main parameters that define this geometry are:

- `fan_angle_deg`: angle of irradiation in the x-y plane.

- `sdd_mm`: Source to Detector Distance in mm.

- `sad_mm`: Source to Axis Distance in mm.

Note that the size of the cone of radiation along the axial (z) direction is equal to the axial extension of the detector.

2.4.8 Serialization/de-serialization of objects

Python provides a module that allows the user to serialize and de-serialize objects: the `pickle`. In DAPHNE you can access `pickle` serialization through the `Pickle` and `Unpickle` utilities, that are included in the `Misc` package. An example of how to serialize and de-serialize an experimental setup is show in listing 2.1. Note that, objects serialized with pickle are not human readable nor are they in a standard format that can be opened with external programs. Therefore, if you plan to save images into a format that can be easily visualized with some standard image software, save it to disk using a Python module such as `imageio`.

```
1 from Misc.Utils import UnPickle,Pickle
2
3 my_experimental_setup = UnPickle("my_file")
4 # set all the parameters
5 my_experimental_setup.Update()
6 Pickle(my_experimental_setup,"my_file2", ".exp")
```

Listing 2.1: Code showing how to load and save an experimental setup from/to file.

2.4.9 Rendering an experimental setup

Once an experimental setup is created and after the `Update` method is called, the rendering of the experimental setup can be achieved using the `Draw()` method.

This will open a 3D interactive scene where the TR, the pixels and sources are displayed. When using a Jupyter notebook a screenshot of the scene can be obtained invoking the method `Draw(use_jupyter=1)`.

2.4.10 3D stack visualization

In order to preview reconstruction results, we developed a 3D image visualizer based on VTK [28]. It can be found in the `Misc`, function `Visualize3dImage`. The `Visualize3dImage` can be used as shown in listing 2.2. Note that the slices of a 3D stack can be browsed using left, right arrow.

```
1  from Misc.Preview import Visualize3dImage
2  ...
3  # display a 3d image as a stack of slices
4  # note that you can navigate through slices using right
        and left arrows
5  Visualize3dImage(img,slice_axis=2)
```

Listing 2.2: Code showing how to display a 3D image as a stack of slices.

## CHAPTER 3

Analytical reconstruction algorithms

This chapter deals with analytical methods for image reconstruction. These methods are based on the assumption that the process of acquiring the projection data of a given object function f, defined as a continuous function in a compact support in $\mathbb{R}^n$, can be modeled through a linear and continuous operator applied to f. The resulting projection data, commonly denoted as p (for parallel beam) or g (in divergent geometries), are in turn continuous functions. Practical algorithms are then obtained by (i) finding an analytical inversion formula for the continuous-space projection function (which can be seen as an integral transform of f), followed by (ii) a discretisation of such an inverted transform. In most cases, data preprocessing steps are introduced in order to compensate for either image noise or data nonlinearity. Even though analytical reconstruction methods don't allow accurate physical modelling, they are the standard methods for CT and still remain indispensable tools in PET. In fact, analytical methods have the big advantage of being less computationally intensive (besides exact 3D methods that are not covered in this book), as compared to iterative ones, and they are generally easy to implement.

The Radon transform (RT), $\mathcal{R}$, (see Section 1.4.2) describes well the idealized continuous-space acquisition process, at least for objects defined in a 2D spatial domain. When working with 3D objects, it is still allowed (and commonly done in practice) to focus on the 2D RT of a discrete sample of 2D slices of f, acquiring and reconstructing them independently, and then composing a 3D object by stacking all the obtained 2D slices. The practical usage of the 3D RT is quite unusual, even though in 3D cone beam geometry (which is relevant for most modern flat-panel based CT systems) the relationship between the set of line integrals (or, more precisely, its radial derivative) and the 3D RT can be exploited (see Grangeat 1987 [10]). The stacked-2D approach to 3D reconstruction plays an important role, real and practical, in image reconstruction. Thus, this approach is not only important for historical reasons. This is true for both CT and PET. In CT, 2D analytical reconstruction is employed both on first generations of multi-detector CT systems (MDCT), as well as in helical CT; in PET, the problem of 3D reconstruction is very often reduced to a set of 2D problems by using specialized techniques to rebin the oblique LOR's on a set of 2D sinograms. For this reason, having a solid understanding of how 2D analytical reconstruction works is of big importance.

This chapter will then focus first on 2D reconstruction, starting from the parallel beam geometry (used directly in PET and in specialized configurations of CT, such as those implemented at synchrotron facilities). Then, the topic of divergent beam, or fan beam 2D reconstruction, will be covered. This is especially important in CT imaging. The extension of 2D fan beam reconstruction to fully 3D reconstruction in cone beam geometry will be described later, due to its importance in modern CT, cone beam CT and micro-CT systems.

3.1 2D RECONSTRUCTION IN PARALLEL BEAM GEOMETRY

All the algorithms presented in this chapter are based on the Central Section Theorem (CST, Section 1.4.5), connecting the Fourier transform (FT) of an object with its RT. For this reason, these methods are also known as Fourier-based methods of image reconstruction. The most straightforward algorithm based on CST is the so-called Direct Fourier Reconstruction (DFR), even though the most used method in practice today is the Filtered Backprojection (FBP), as described later on this section.

3.1.1 Direct Fourier Reconstruction (DFR)

Let $F = \mathcal{F}_2 f$, be the 2D FT of the object f. If F is known, we can reconstruct f by just taking its inverse 2D FT, i.e., $f = \mathcal{F}_2^{-1} F$. We have already shown that the CST (Equation 1.6) provides a direct link between the projections of f and F in signed polar coordinates, as $P(\nu, \phi) \overset{\text{CST}}{=} F(u', 0) = F(\nu, \phi)$. The previous identity can be intuitively exploited to formulate a reconstruction algorithm, called Direct Fourier Reconstruction (DFR) and reported below as Algorithm 1:

Algorithm 1 - DFR

---

1. Take the 1D FT of p, $P = \mathcal{F}_1 p$, for all the available projections taken at angles ϕ_i;

2. exploit the CST by taking samples of the function F in the frequency domain from the results of point 1, thus obtaining an approximated version $\widetilde{F}$ of the 2D FT of f;

3. reconstruct the image by taking the inverse 2D FT of $\widetilde{F}$, i.e., $\widetilde{f}_{\text{DFR}} = \mathcal{F}_2^{-1} \widetilde{F}$.

This reconstruction method is attractive for its conceptual simplicity, and also because it can take advantage of the very efficient techniques available for the computation of the discrete FT, or DFT (and of its inverse), known as fast Fourier transform or FFT. From a practical point of view, the data in a real PET or CT acquisition is only available as an array of discrete samples, with sampling step $\Delta x'$ in the spatial domain and $\Delta \nu = 1/\Delta x'$ in the frequency domain. That is, the available samples of F are arranged in a polar grid as shown in Figure 3.1.

Taking the (discrete) inverse 2D FT would require us to have F sampled on a rectangular grid, which implies an interpolation procedure, also referred to as "gridding" in this context. Unfortunately, interpolation in the frequency domain is generally a hard task as it leads to image artifacts when the inverse FT is performed to convert data back to the spatial domain. Hence, more advanced polar-to-Cartesian interpolation strategies rather than simple linear or bi-linear interpolation are employed.

In what follows, we will provide a possible Python implementation of Algorithm 1. This implementation is

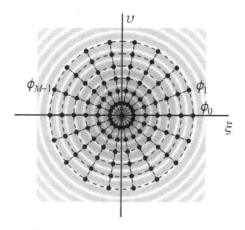

Figure 3.1: As a consequence of the central section theorem, the available samples of F are arranged in a radial grind in the frequency space.

available in the Jupyter notebook DFRDemo that can be found on the companion software, and was inspired from Ref. [3]. It requires a few functions from the *Numpy* and *Scipy* packages. Let us start with the following imports:

```
 1 import numpy as np
 2 import matplotlib
 3 import matplotlib.pyplot as plt
 4
 5 import scipy.interpolate
 6 import scipy.misc
 7 import scipy.ndimage.interpolation
 8
 9 N_ang = 180 # Number of angular samples
10 N_rad = 256 # Number of radial samples
11 Imsize_X = 256 # Phantom size on x (pixel)
12 Imsize_Y = 256 # Phantom size on y (pixel)
```

Listing 3.1: DFR example, part 1 (importing modules and setting global variables).

where we have also set some global parameter regarding the size of the image and the sinogram to be imported in this example. Let us now load the Shepp-Logan phantom and its parallel beam sinogram:

```
14 # Load phantom data and sinogram data from disk
15 phantom = np.fromfile("../Data/SL_HC_256x256.raw",dtype=
        np.float32).reshape((Imsize_Y,Imsize_X))
16 sino = np.fromfile("../Data/SL_HC_180x256_paralsino.raw"
        ,dtype=np.float32).reshape((N_rad,N_ang))
17
18 # Plot phantom and sinogram
19 plt.figure(figsize=(10,5))
20 plt.subplot(121)
21 plt.title("Phantom")
22 plt.imshow(phantom)
23 plt.subplot(122)
24 plt.title("Sinogram")
25 plt.imshow(sino)
26 plt.tight_layout()
```

Listing 3.2: DFR example, part 2 (loading and displaying phantom data and sinogram).

The output is shown in Figure 3.2. Now, the implementation of the DFR requires the computation of the 1D FFT of the sinogram along the radial direction (i.e., along x').

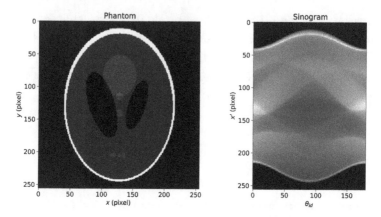

Figure 3.2: Shepp-Logan head phantom and its parallel beam sinogram.

It is instructive also to take the 2D FFT of the phantom itself, just for comparison:

```
28  # Take the 1D FFT of the sinogram data along radial
         direction
29  sino_fft1d=np.fft.fftshift(
30              np.fft.fft(
31                  np.fft.ifftshift(
32                      sino,
33                      axes=0
34                  ),
35              axis=0
36              ),
37          axes=0
38          )
39
40  # Compute the 2D FFT of the phantom
41  phantom_fft2d=np.fft.fftshift(
42              np.fft.fft2(
43                  np.fft.ifftshift(
44                      phantom
45                  )
46              )
47          )
```

Listing 3.3: DFR example, part 3 (calculate 1D FFT of the sinogram and 2D FFT of the phantom).

The usage of the *Numpy* functions **fftshift** and **ifftshift** in the above listing is required to have the correct offset of the transformed datasets in the frequency domain.

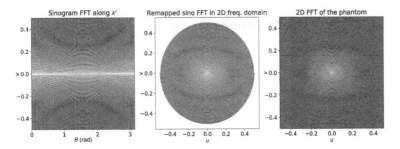

Figure 3.3: Application of the CST to the sinogram data of Figure 3.2, and comparison with the 2D FFT of the object. This figure is the result of the code listing 3.4.

We can now exploit the Central Section Theorem by remapping the 1D FFT of the sinogram in the 2D frequency space. First, let us create the values of the polar coordinates of each sample in the FFT transformed sinogram, along with the corresponding values in cartesian coordinates. The result is shown in Figure 3.3.

```
49  # Coordinates of the Fourier transformed sinogram
50  th=np.array([np.pi*i/N_ang for i in range(N_ang)])
51  nu=np.arange(N_rad)-N_rad/2
52  th,nu=np.meshgrid(th,nu)
53  nu=nu.flatten()
54  th=th.flatten()
55
56  # Coordinates of 2D frequency space
57  u_pol=(N_rad/2)+nu*np.cos(th)
58  v_pol=(N_rad/2)+nu*np.sin(th)
59
60  # Display the 1D radial FFT of the sinogram
61  plt.gray()
62  plt.figure(figsize=(16,5))
63  plt.subplot(131)
64  plt.title(r"Sinogram FFT along $x^{\prime}$")
65  plt.xlabel(r"$\theta$ (rad)", labelpad=0)
66  plt.ylabel(r"$\nu$", labelpad=-10)
67  plt.imshow(np.log(np.abs(sino_fft1d)), extent=[0,np.pi
        ,-0.5,0.5], aspect="auto")
68
69  # Display the remapped 1D FFT of the sinogram on 2D
        frequency space
70  plt.subplot(132)
71  plt.title("Remapped sino FFT in 2D freq. domain")
72  plt.scatter(
73      (u_pol-Imsize_X/2)/Imsize_X,
74      (v_pol-Imsize_Y/2)/Imsize_Y,
```

```
75        c=np.log(np.abs(sino_fft1d.flatten())),
76     marker='.',
77     edgecolor='none'
78     )
79 plt.xlabel(r"$u$")
80 plt.ylabel(r"$v$", labelpad=-10)
81
82 # Display the 2D FFT of phantom for comparison
83 plt.subplot(133)
84 plt.title("2D FFT of the phantom")
85 plt.imshow(np.log(np.abs(phantom_fft2d)), extent
        =[0-0.5,0.5,-0.5,0.5], aspect="auto")
86 plt.xlabel(r"$u$")
87 plt.ylabel(r"$v$", labelpad=-10)
```

Listing 3.4: DFR example, part 4 (display the 1D FFT of the sinogram, its remapping in 2D frequency space (as per the CST) and 2D FFT of the phantom).

Finally, we can create an interpolated version of the 2D FFT of the object from just the samples of the 1D Fourier transformed sinogram using the following code, employing the **griddata** function of the **scipy.interpolate** module of *SciPy*:

```
89 # Create a rectangular grid of points in 2D frequency
        domain
90 u_cart,v_cart=np.meshgrid(np.arange(N_rad),np.arange(
        N_rad))
91 u_cart=u_cart.flatten()
92 v_cart=v_cart.flatten()
93
94 # Interpolate the 2D Fourier space grid from the
        transformed sinogram
95 remapped_fft2d=scipy.interpolate.griddata(
96        (u_pol,v_pol),
97        sino_fft1d.flatten(),
98        (u_cart,v_cart),
99        method='cubic',
100        fill_value=0.0
101     ).reshape((N_rad,N_rad))
```

Listing 3.5: DFR example, part 5 (interpolate the 1D FFT of the sinogram in cartesian coordinates).

After that, the only (straightforward) remaining step is the calculation of the 2D IFFT of the interpolated FFT data. The following listing shows how to do this with the **np.fft** package of *Numpy*, and how to display

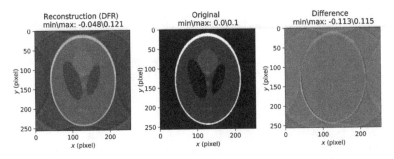

Figure 3.4: DFR reconstruction of the Shepp-Logan head phantom, obtained by direct application of the CST and cubic interpolation in the frequency domain.

the result with matplotlib. The final result is shown in Figure 3.4.

```
103 # The final step - reconstruct the image
104 # Inverse transform from 2D FFT to the image space
105 recon=np.real(
106     np.fft.fftshift(
107         np.fft.ifft2(
108             np.fft.ifftshift(remapped_fft2d)
109         )
110     )
111 )
112
113 # Compute the error image (original - recon)
114 difference=np.subtract(phantom,recon)
```

Listing 3.6: DFR example, part 6 (reconstruct image by DFR and compute the error image).

The remaining code of this example just shows how to display the result. Expert scientific Python programmers might find this listing useless, but we preferred to add this for all those who are still not very familiar with matplotlib and related functionalities.

```
116 # Display the reconstructed image and difference with
        original
117 plt.figure(figsize=(16,7))
118 plt.subplot(131)
119 plt.title("Reconstruction \n min\max: " + str(np.around(
        np.min(recon),3)) + "\\" + str(np.around(np.max(
        recon),3)))
120 plt.imshow(recon, interpolation='bicubic')
```

```
121 plt.xlabel(r"$x$ (pixel)")
122 plt.ylabel(r"$y$ (pixel)", labelpad = 0)
123 plt.subplot(132)
124 plt.title("Original \n min\max: " + str(np.around(np.min
        (phantom),3)) + "\\" + str(np.around(np.max(phantom)
        ,3)))
125 plt.imshow(phantom, interpolation='bicubic')
126 plt.xlabel(r"$x$ (pixel)")
127 plt.ylabel(r"$y$ (pixel)", labelpad = 0)
128 plt.subplot(133)
129 plt.title("Difference \n min\max: " + str(np.around(np.
        min(difference),3)) + "\\" + str(np.around(np.max(
        difference),3)))
130 plt.imshow(difference, interpolation='bicubic')
131 plt.xlabel(r"$x$ (pixel)")
132 plt.ylabel(r"$y$ (pixel)", labelpad = 0)
133 plt.tight_layout()
```

Listing 3.7: DFR example, part 7 (display the result and error image).

As it appears from the reconstructed image in Figure 3.4, the direct application of the CST to the DFR reconstruction yielded a relatively poor result, with several artifacts. Indeed, re-gridding of Fourier-space data with non-uniform sampling is far from being an easy task. This issue is usually addressed by pre-filtering the available FFT samples by means of suitable kernels (e.g., Kaiser-Bessel), as well as compensation of variable sample density across the explored frequency space. This is especially useful for reconstructing images in Magnetic Resonance Imaging (MRI), where the k-space filling trajectory is different depending on the acquisition strategy. In CT/PET reconstruction, the typical radial filling of the frequency space from projection data had led the community to employ different reconstruction strategies, such as the Filtered Backprojection which is described in the following sections.

3.1.2 Filtered Backprojection (FBP)

A different reconstruction formula arises by rewriting the inverse 2D FT of F as follows:

$$f(x,y) = \int\limits_{-\infty}^{\infty} \mathrm{d}v \int\limits_{-\infty}^{\infty} \mathrm{d}\xi \, F(\xi, v) e^{j2\pi(\xi x + vy)}$$

$$= \int\limits_{0}^{\pi} \mathrm{d}\phi \int\limits_{-\infty}^{\infty} \mathrm{d}\nu \, F(\nu, \phi) |\nu| e^{j2\pi\nu(x\cos\phi + y\sin\phi)}$$

$$\overset{\mathrm{CST}}{=} \int\limits_{0}^{\pi} \mathrm{d}\phi \int\limits_{-\infty}^{\infty} \mathrm{d}\nu \, P(\nu, \phi) |\nu| e^{j2\pi\nu(x\cos\phi + y\sin\phi)} \quad (3.1)$$

where ξ, v are the coordinates of the 2D frequency domain, $\mathrm{d}\xi \mathrm{d}v = |\nu| \mathrm{d}\nu \mathrm{d}\phi$, and $|\nu|$ is the Jacobian determinant of the cartesian to signed cylindrical coordinates transformation. Readers with a basic background in the theory of signal processing should figure out immediately that the inner integral in Equation 3.1 is equivalent to a 1D filtering of the sinogram $p(x', \phi)$ along the radial variable x', with a filter with frequency response equal to $|\nu|$. By applying the convolution theorem, this equation can be rewritten as:

$$f(x,y) = \int\limits_{0}^{\pi} \mathrm{d}\phi \int\limits_{-\infty}^{\infty} \mathrm{d}x' \, p(x', \phi) h(x\cos\phi + y\sin\phi - x')$$

$$(3.2)$$

Basically, we have exploited the fact that the inverse 1D FT of the product $P(\nu, \phi)|\nu|$ is equal (by the convolution theorem) to the result of the 1D convolution $p(x', \phi) * h(x')$. In this notation, h is a generalized function expressing the so-called "ramp filter" due to its shape in the frequency domain:

$$h(x') = \int\limits_{-\infty}^{\infty} \mathrm{d}\nu \, |\nu| \, e^{j2\pi\nu x'} \quad (3.3)$$

In the jargon of signal processing, the function h is called the kernel of the ramp filter (or, equivalently, its impulse response). In a more compact form, we can rewrite Equation 3.2 as

$$
\begin{aligned}
f(x,y) &= \int_0^\pi \mathrm{d}\phi \, (p * h) \, (x\cos\phi + y\sin\phi, \phi) \\
&= \int_0^\pi \mathrm{d}\phi \, q(x\cos\phi + y\sin\phi, \phi)
\end{aligned}
\tag{3.4}
$$

where we have defined the auxiliary function $q = p * h = \mathcal{F}_1^{-1}(P|\nu|)$. Here and in what follows, we will refer to the function q as the filtered sinogram.

3.1.2.1 Filtered Backprojection vs. Convolution Backprojection

The inversion formula just written leads to a reconstruction algorithm called Filtered Backprojection (FBP). This name is better suited when the inner integral is performed in the frequency space (see Equation 3.1); if instead the inner integral is written in the form of a convolution in the space domain, as in Equations 3.2 and 3.4, then the same algorithm is conventionally called Convolution-Backprojection (CBP). From a practical point of view, filtering of 1D signals in the frequency domain is more efficient than in the space domain, because computer implementations of this task can take advantage of very efficient strategies for the calculation of forward and backward FFT (both using general purpose central processing units or CPU's, hardware-based tools such as Field Programmable Gate Arrays or FPGA's or, more recently, also graphic processing units or GPU's).

3.1.2.2 Ramp filter and apodisation windows

As the name suggest, the (2D) FBP algorithm is made up of two basic "blocks": a 1D filtering (the inner integral of Equations 3.1-3.4), and a 2D backprojection (the outer integral in the same equations). It must be noted that the integral definition of the ramp filter kernel h in Equation 3.3 is meaningful only in the sense of distributions. In fact, the function $|\nu|$ is not L^1-integrable and thus it does not admit an inverse FT. Nevertheless, an explicit expression for the ramp filter impulse response h can be found in practice by assuming that the object function f is band-limited, i.e., $F(\nu, \phi) = 0$ for $|\nu| > \nu_{max}$. If f is band-limited, it follows from the CST that $P(\nu, \phi) = 0$ for $|\nu| > \nu_{max}$ (in other words, also p is band-limited) and hence the inner integral in Equation 3.4 can be restricted to the compact interval $[-\nu_{max}; \nu_{max}]$ without loss of generality. A natural choice for the limiting frequency can be made by remembering that, in practice, all sinograms are sampled with step $\Delta x'$ and hence we can set $\nu_{max} = \nu_{Nyquist} = 1/(2\Delta x')$. In practical applications, a further modification of the ramp filter is done by means of apodisation windows, that we will generically denote as $A(\nu)$ in this context:

$$
\tilde{h}(x') = \int_{-\nu_{max}}^{\nu_{max}} d\nu \, A(\nu) \, |\nu| \, e^{j2\pi\nu x'}
$$

$$
= \int_{-\infty}^{\infty} d\nu \, \Pi\left(\frac{\nu}{2\nu_{max}}\right) A(\nu) \, |\nu| \, e^{j2\pi\nu x'} .
$$

(3.5)

where Π is the boxcar (or rect) function.

Kak and Slaney [14] reported an analytical expression of $\tilde{h}$ in the simplest case of flat apodisation window (i.e., $A(\nu) = 1$) with cut-off frequency equal to $1/\Delta x'$ (this is also known as Ramachandran-Lakshminarayan filter, or, shortened, Ram-Lak filter):

$$
\tilde{h}(x') = \frac{1}{(2\Delta x')^2} \left[2\mathrm{sinc}\left(\frac{x'}{\Delta x'}\right) - \mathrm{sinc}^2\left(\frac{x'}{2\Delta x'}\right) \right],
$$

(3.6)

with $\text{sinc}(x) = \sin(\pi x)/(\pi x)$ the cardinal sine function and $\Delta x'$ the sampling step, linked to the limiting frequency $\nu_{\max}$ by the relation

$$\Delta x' = \frac{1}{2\nu_{\max}} . \tag{3.7}$$

The discrete version of the filter impulse response in Equation 3.6 is

$$\tilde{h}(k\Delta x') = \frac{1}{(2\Delta x')^2} \left[2\text{sinc}\,(k) - \text{sinc}^2\,(k/2) \right]$$

$$= \frac{1}{(2\Delta x')^2} \begin{cases} 1 & \text{if } k = 0, \\ 0 & \text{if } k \text{ is even,} \\ -1/(\pi k/2)^2 & \text{if } k \text{ is odd.} \end{cases} \tag{3.8}$$

The IR and FR of the Ram-Lak filter are shown in Figure 3.5 below. Within our reconstruction framework, DAPHNE, the generation of the Ram-Lak filter is done by the `GenerateRamp()` function of the **FBP** class. Let us report this function here:

```
12    def GenerateRamp(self, tau=1):
13        """!@brief
14            Generate and return a ramp filter
15        """
16        filter_size = self._sinogram._radial_bins
17        self._h = np.zeros((filter_size))
18        # find all the indices odd with respect to
          filter_size/2
19        idx = np.arange(1, filter_size, 2) + filter_size
          % 2
20        self._h[idx] = -(((idx - filter_size // 2) * tau
          * np.pi) ** 2)
21        nnull = self._h != 0
22        # generate the filter impulse response
23        self._h[nnull] = 1 / self._h[nnull]
24        self._h[filter_size // 2] = 1 / (4 * (tau) ** 2)
25        # take the FFT of the filter IR to generate its
          frequency response
26        ramp = np.fft.fft(self._h) * self._sinogram.
          _radial_step_mm / 2
27        H = abs(ramp)
28        # this creates a matrix where each column
          contains H
29        # (useful to accelerate the row-by-row
          multiplication of the sinogram FFT)
```

```
30      self._Hm = np.array([H] * self._sinogram.
    _angular_bins)
31      self._Hm = np.transpose(self._Hm)
32      self._Hm = np.tile(self._Hm,self._sinogram.
    _z_bins).reshape(self._sinogram._radial_bins,self.
    _sinogram._angular_bins,-1)
```

Listing 3.8: Function generating a 3D matrix containing $N_{\mathrm{ang}} \cdot N_{slices}$ repetitions of the Ram-Lak filter 1D frequency response $H(\nu)$. This function is implemented in the FBP class of the DAPHNE framework.

The parameter tau is just the sampling step $\Delta x'$, by borrowing the notation of Ref. [14]. The function is such that the spatial domain impulse response (IR) has its origin at filter_size // 2, where // is the floor division operator (see the Python language reference).

In the last step, we have just converted the 1D array H to a 3D array, denoted by self._Hm, which contains nothing but the repetition of the same filter frequency response H for a number of times equal to self._sinogram._angular_bins * self._sinogram.z_bins. This approach helps us to implement the application of the ramp filter to the sinogram in the frequency space as just an element-wise multiplication of two matrices (the filter matrix and the sinogram FFT matrix). The filtering step is implemented in the FilterSino() function of the FBP class, using the fft package of the *Numpy* library:

```
1 # perform FFT of the sinogram along x'
2 self.fft1d_sinogram = np.fft.fft(self._sinogram._data,
    axis=0)
3 # apply filter and perform 1D IFFT
4 self._filtered_sinogram = np.real(
5     np.fft.ifft(
6         self.fft1d_sinogram * self._Hm, axis=0
7     )
8 )
```

Listing 3.9: Ramp filtering of the sinogram in the frequency domain.

It is now useful to compare the filtered sinogram with the unfiltered one. The listing 3.10 on page 43 will show the two sinograms, p and $\tilde{q}$ along with a line profile

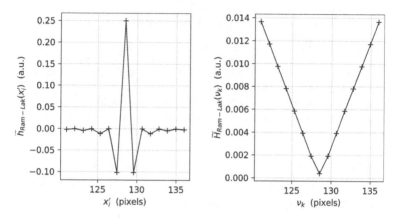

Figure 3.5: Ram-Lak (ramp) filter IR (left) and FR (right), generated by the `GenerateRamp()` function of the `FBP` class. The plot range was reduced with respect to the original filter size (256 samples in this example) to increase the visibility of the relevant portion of the filter IR and FR. Note the value > 0 of the filter FR at DC (k=128) (see Ref. [14] for more details).

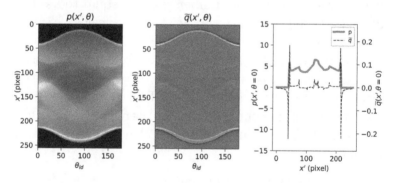

Figure 3.6: Output of the listing 3.10. Note the negative values appearing in the filtered sinogram.

of both sinograms at $\theta = 0$. The output is shown in Figure 3.6.

The notation used in Equations 3.5 and 3.6 emphasises the conceptual difference between the real filter kernel $\tilde{h}$ and the generalised function h. In most workstations, the type of window A is one of the (indeed, very few) parameters of image reconstruction that the user is allowed to change when using FBP. Briefly, we will just say that such windows allow the user to modify the shape of the ramp filter by attenuating the high spatial frequency and then reducing the noise in the final image.

We can now write an inversion formula for p, which is the basis for all practical implementations of the FBP algorithm:

$$
\begin{aligned}
\widetilde{f}_{\text{FBP}}(x, y) &= \int_0^{\pi} \mathrm{d}\phi \int_{-\infty}^{\infty} \mathrm{d}x'\, p(x', \phi)\widetilde{h}(x\cos\phi + y\sin\phi - x') \\
&= \int_0^{\pi} \mathrm{d}\phi \left(p * \widetilde{h}\right)(x\cos\phi + y\sin\phi, \phi) = \\
&= \int_0^{\pi} \mathrm{d}\phi\, \widetilde{q}(x\cos\phi + y\sin\phi, \phi)
\end{aligned}
$$

$$(3.9)$$

with obvious meaning of the modified filtered sinogram $\widetilde{q} = p * \widetilde{h}$.

```
1  # Subplot 1 - Original sinogram
2  fig = plt.figure(figsize=(8,3.5))
3  plt.subplot("131")
4  plt.imshow(f.sinogram.data[:,:,0])
5  plt.title(r"$p(x^\prime,\theta)$")
6  plt.xlabel(r"$\theta_{id}$")
7  plt.ylabel(r"$x^{\prime}$ (pixel)", labelpad=-2)
8
9  # Subplot 2 - Filtered sinogram
10 plt.subplot("132")
11 plt.imshow(f.filtered_sinogram[:,:,0])
12 plt.title(r"$\widetilde{q}(x^\prime,\theta)$")
13 plt.xlabel(r"$\theta_{id}$")
14 plt.ylabel(r"$x^{\prime}$ (pixel)", labelpad=-2)
15
16 # Subplot 3 - Line profiles of original and filtered
       sinograms
17 ax1 = plt.subplot("133")
18 l1 = ax1.plot(f.sinogram.data[:,1],
19         linewidth=3, color='gray', label="p")
20 ax1.set_ylim([-15,15])
21 ax1.set_xlabel(r"$x^{\prime}$ (pixel)")
22 ax1.set_ylabel(r"$p(x^\prime,\theta=0$)",
23              labelpad=-7)
```

```
24
25 ax2 = ax1.twinx()
26 l2 = ax2.plot(f.filtered_sinogram[:,1],
27         linewidth=1, linestyle='dashed', color='black',
28           label=r"$\widetilde{q}$")
29 ax2.set_ylim([-0.275,0.275])
30 ax2.set_ylabel(r"$\widetilde{q}(x^\prime$,$\theta=0$)",
31           labelpad=-3)
32
33 # Create legend
34 lns = l1+l2
35 labs = [l.get_label() for l in lns]
36 ax1.legend(lns, labs, loc=0, prop={'size': 8})
37
38 plt.tight_layout()
39 plt.subplots_adjust(wspace=0.4)
```

Listing 3.10: Display the unfiltered vs. ramp-filtered sinogram of the Shepp-Logan phantom.

3.1.2.3   The backprojection step

The integration with respect to the angular variable in the equations above in this section is called "backprojection" (BP), representing the second step of the FBP algorithm which follows the application of the ramp filter to each radial section of the sinogram. We will denote the backprojection operator by the symbol $\mathcal{R}_n^\#$, defined in two dimensions by means of the identity

$$f = \mathcal{R}_2^\# q . \qquad (3.10)$$

By comparing Equations 3.10 and 3.2, it appears clear that the backprojection is not the inverse of the projection operator, or Radon transform, because in Equation 3.10 we applied $\mathcal{R}_2^\#$ to q (i.e., the filtered sinogram) instead of $p = \mathcal{R}_2 f$ by obtaining f. We could then write

$$f = \mathcal{R}_2^\# (\mathcal{R}_2 f * h) . \qquad (3.11)$$

as an equivalent expression of Equation 3.2. From a practical point of view, backprojecting is similar to smearing back each profile on the image plane along its original direction of projection. Each backprojected profile will give rise to an intermediate image.

A possible workflow for a practical implementation of the FBP algorithm is reported in Algorithm 2 on page 45.

Algorithm 2 - FBP, 2D PARALLEL BEAM

---

1. Select the apodisation window $A(\nu)$ and compute the discrete version of the modified filter kernel $\tilde{h}$;

2. apply the 1D ramp-filter to the sinogram: for each available projection angle ϕ, take the 1D DFT's P and $\tilde{H}$ of the sinogram and of the modified filter, respectively, and multiply them in the frequency domain; afterwards, compute the filtered sinogram $\tilde{q}$ as the inverse 1D DFT of the product $P \cdot \tilde{H}$, i.e., $\tilde{q} = \mathcal{F}_1^{-1}(P \cdot \tilde{H})$;

3. reconstruct the image by backprojecting each row of the filtered sinogram $\tilde{q}$ on the image plane along its own direction of projection ϕ, i.e., $\tilde{f}_{\text{FBP}} = \mathcal{R}_2^{\#} \tilde{q}$.

Implementing the backprojection step in Python requires the import of some function for data interpolation. We will import `interp1d` from `scipy.interpolation` to this purpose. The code reported in listings 3.11, 3.12 and 3.13 will produce an intermediate reconstruction $\tilde{f}$ by backprojecting just one filtered projection. The output is shown in Figure 3.7. As a first step we need to initialize the target image (`img_partial`) as well as some auxiliary variables:

```
1  # Create a squared target image
2  Imsize = Imsize_X
3  img_partial = np.zeros([Imsize,Imsize], dtype=float)
4
5  # Filtered sinogram just copied here
6  s_filt=f._filtered_sinogram[:,:,0]
7
8  # Grid of the coordinates of the pixel centers
9  # in the target image (pitch=1 for simplicity)
10 x,y = np.mgrid[:Imsize, :Imsize] - Imsize // 2 + 0.5
11
12 # Create an array with the radial coordinates
13 # of the detector pixels
14 xp_det = np.arange(N_rad) - N_rad // 2 + 0.5
```

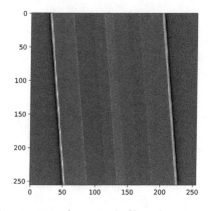

Figure 3.7: Backprojection of one single filtered projection of the Shepp-Logan phantom, as resulted from listings 3.11, 3.12 and 3.13 (pages 45-47.

```
15
16 # Create an array with the gantry angles
17 theta_deg = np.arange(0, N_ang, 1)
```

Listing 3.11: Initialisation of data structures and variables for the partial backprojection example.

Afterwards, we select a specified projection from the filtered sinogram to be backprojected into the destination image. In this example, we have selected the projection taken at $\theta = 5$ degrees.

```
1 # Select an angle of projection from its index
2 # (let's take id=5, which means theta_deg=5 degrees
3 # in this example)
4 id_proj = 5
5
6 # This tuple stores the projection (1D array)
7 # at the selected angle and the corresponding
8 # gantry angle in radians
9 (proj, theta_rad) = (s_filt.T[id_proj,:],
10                      np.deg2rad(theta_deg)[id_proj])
11
12 # This holds the values of x' (in gantry reference frame
      )
13 # of each pixel
14 xp = y * np.cos(theta_rad) - x * np.sin(theta_rad)
```

Listing 3.12: Selection of the filtered projection for the partial backprojection example.

Finally, we can import and use the *Scipy* function
`interp1d()` to backproject the selected filtered projection into `img_partial`:

```
1  # Import the required scipy function for interpolation
2  from scipy.interpolate import interp1d
3
4  # Backproject
5  img_partial += interp1d(
6      x = xp_det,
7      y = proj,
8      kind = 'linear',
9      bounds_error = False,
10     fill_value = 0,
11     assume_sorted = False
12 ) (xp)
13
14 plt.figure()
15 plt.imshow(img_partial)
```

Listing 3.13: Backprojection with only one filtered projection, taken at $\theta = 5$ degrees in this example.

In order to compose a full reconstruction, we need to iterate over all acquired projections available in the filtered sinogram. To this purpose, we will just include the accumulation step in a **for** loop, iterating over all tuples (`proj`, `theta_rad`) using the **zip** function:

```
1  img_fbp = np.zeros([Imsize,Imsize], dtype=float)
2
3  for proj, theta_rad in zip(s_filt.T, np.deg2rad(
       theta_deg)):
4      xp = y * np.cos(theta_rad) - x * np.sin(theta_rad)
5
6      img_fbp += interp1d(
7          x = xp_det,
8          y = proj,
9          kind = 'linear',
10         bounds_error = False,
11         fill_value = 0,
12         assume_sorted = False
13     ) (xp)
```

Listing 3.14: Iteration of the backprojection step over all angles of the filtered sinogram.

It is instructive to see the effect of backprojecting the unfiltered sinogram instead of the filtered one, as shown

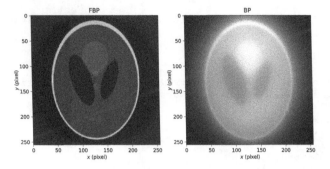

Figure 3.8: FBP vs BP. This results is another empirical demonstration that the backprojection is not the inverse operator of the Radon transform.

in Figure 3.8. This is as easy as changing line 6 of the listing 3.11 by assigning `f.sinogram.data` to the auxiliary array `s_filt` instead of `f.filtered_sinogram`. One problem with this approach is that, if we skip the ramp-filtering step, the backprojected profiles will never superimpose in a destructive way because the projection p is positive definite (as it comes from the line integrals of a positive definite physical quantity, μ). Hence, by BP alone we cannot avoid the "tails" of the reconstructed image away from the object's support.

The ultimate reason why BP alone isn't able to reconstruct the original object, apart for the insufficient number of views in this example, is more understandable by looking at the radial grid in the frequency space shown in Figure 3.1. The sparsity of the samples of the object's 2D FT at high spatial frequency with respect to those at lower spatial frequency has to be compensated for in some way. High frequency samples of the spectrum must be weighted more than low frequency ones in order to compensate for such difference in sampling density. In other words, a filter must be applied to the data before backprojection. Because the sample density decreases linearly with $|\nu|$, the ramp filter appearing in Equation 3.1 is needed to counterbalance the undersampling at high frequencies.

All the Python source code related to this section on parallel-beam FBP can be found in the Jupyter notebook FBPParallelBeamDemo of the companion software. The following example will show instead how to perform the FBP reconstruction within DAPHNE, using the high-level implementation.

3.1.3   High-level Python implementation of the FBP

We will now provide a short overview of how to perform FBP on parallel beam projection data within the DAPHNE framework. The example is done considering a CT acquisition, but it can work with PET data too if a sinogram (or a stack of them) is provided.

Using our library, the following modules must be imported in order to perform the reconstruction:

```
1 from Algorithms.SinogramGenerator import
     SinogramGenerator
2 from Algorithms.FBP import FBP
3 from Geometry.ExperimentalSetupCT import
     ExperimentalSetupCT,Mode,DetectorShape
```

Listing 3.15: Modules required to perform FBP reconstruction.

The geometry must be set up as explained in Chapter 2, by creating an instance of the class implementing the experimental setup for CT. This can be done as simply as adding a line such as: my_experimental_setup = ExperimentalSetupCT(). All the relevant parameters (number of radial and angular bins, detector pitch, etc.) must be manually set in my_experimental_setup.

An instance of the Sinogram class also must be created, on top of the geometric setup available in my_experimental_setup. This is done as follows:

```
1 # Set image size and number of radial bins
2 Imsize_Y = Imsize_X = N_rad = my_experimental_setup.
     _number_of_pixels_per_slice
3 # Set number of angles
4 N_ang = my_experimental_setup._number_of_gantry_angles
```

```
5
6  # Read sino data
7  sino_data = np.fromfile("../Data/
       SL_HC_180x256_paralsino_theta_r.raw",dtype=np.
       float32).reshape((N_rad,N_ang,1))
8
9  # Create an instance of the SinogramGenerator class
10 sino_generator=SinogramGenerator(my_experimental_setup)
11
12 # Generate a sinogram (geometry defined in
       my_experimental_setup)
13 sino=sino_generator.GenerateEmptySinogram()
14
15 # Assign sinogram data to the sino object
16 sino._data=sino_data
```

Listing 3.16: Create a sinogram object and assign the sinogram data to it, based on the geometry of my_experimental_setup.

Now, performing the FBP reconstruction of the Shepp-Logan phantom shown in our example is as easy as writing the following code (listing 3.17).

```
1  # Create an instance of the FBP class
2  f=FBP()
3
4  # Assign the sinogram to the FBP object
5  f._sinogram=sino
6
7  # this is the interpolation for the backprojection
8  # available options are : "linear","nearest","zero","
       slinear","quadratic","cubic"
9  # see for https://docs.scipy.org/doc/scipy/reference/
       generated/szerocipy.interpolate.interp1d.html
       parameter: kind
10 # for an explanation of the interpolation parameters
11 f._interpolator='cubic'
12
13 # Perform reconstruction
14 f.Reconstruct()
```

Listing 3.17: Perform the FBP reconstruction of the sino object, containing the projection data shown in the examples above and with the geometry implemented in my_experimental_setup.

A possible output of this code is shown Figure 3.9, where cubic interpolation has been used in the backprojection step.

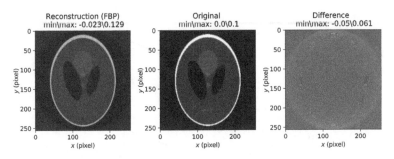

Figure 3.9: FBP reconstruction of the Shepp-Logan head phantom, compared to the original image and the difference image.

3.2 2D FBP IN FAN BEAM GEOMETRY

We will now extend the results obtained in the previous section in order to reconstruct images from projections acquired with divergent beams. This section and the following one concerning the cone beam geometry are primarily of interest for CT imaging. In fact, most of today's medical CT scanners produce divergent projections. The 2D fan beam geometry and related parameters are shown in Figure 3.10.

The practical implementation of this geometry can rely on both flat or curved detectors: in this section, we will just consider the most common cases of centered flat detectors (especially useful when flat-panel X-ray detectors are used) and circular arc detectors (which is the standard for clinical CT scanners).

3.2.1 Rebinning

For a 1st or 2nd generation scanner operating in parallel beam geometry, the acquired data was easily linked to the 2D RT of the object as all the line integrals were collected sequentially on the radial (x') and then on the angular (ϕ) direction.

In a 3rd generation scanner, each projection at a given gantry angle contains data from several angles ϕ and

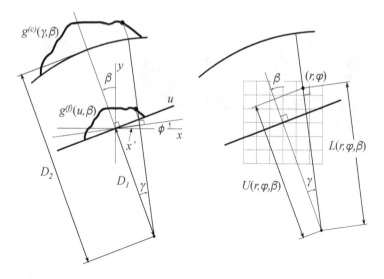

Figure 3.10: Fan beam geometry and meaning of the related coordinates and variables.

hence the fan beam sinogram that is formed by stacking all the projections taken at all the gantry angles does not coincide with the Radon transform of f. On the other hand, it is intuitive that all the data needed to "build" the 2D RT of f is somehow included in the fan-beam data. Resampling, or rebinning, the fan-beam data into a parallel beam dataset is one possible solution of the reconstruction problem in fan beam geometry.

Let us now denote by β the gantry angle, i.e., the angle between the central ray of the beam and the y-axis, and let us denote by γ the angle between each detected ray at a given gantry angle and the central ray (see Figure 3.10). The detector (radial) coordinate on a flat array is denoted by u. For this type of detector, all the reconstruction formulae are obtained by ideally placing the detector array in such a way that it passes from the axis of rotation (i.e., leading to magnification $m = 1$). We will refer to it as a virtual detector. The parallel beam data can be calculated from the fan beam data $g^{(f)}$ or $g^{(c)}$ (in case of flat or curved detector, respectively) observing that

$$p(x', \phi) = g^{(f)}(u(x'), \beta(x', \phi))$$
$$= g^{(c)}(\gamma(x', \phi), \beta(x', \phi)),$$

(3.12)

where the fan beam coordinates can be calculated for each desired bin of the Radon space with the following formulae:

$$u(x') = \frac{D_1 x'}{\sqrt{D_1^2 - x'^2}},$$

$$\gamma(x') = -\sin^{-1} \frac{x'}{D_1},$$

$$\beta(x', \phi) = \phi + \gamma(x').$$

(3.13)

The rebinning is done by filling up the parallel beam sinogram p by taking, for each position of the x', ϕ space, the corresponding line integral value from the fan beam sinogram. Afterwards, reconstruction is just done by the 2D FBP for parallel beam geometry. From a practical point of view, mapping the (u, β) coordinates or the (γ, β) coordinates to the Radon coordinates requires some type of interpolation.

3.2.2 Full-scan (2π) FBP reconstruction in native fan beam geometry

Another approach to the reconstruction of fan beam data is to adapt the FBP formula to the specific geometry under consideration. The derivation reported here is based on the ones of Kak and Slaney [14] and Turbell [31], based on the results of Herman and Naparstek [11]. Let us first rewrite the FBP reconstruction formula by expressing the image point location in (unsigned) cylindrical coordinates, (r, φ), where $r \in [0, \infty[$ and $\varphi \in [0, 2\pi[$:

$$f(r, \varphi) = \frac{1}{2} \int\limits_0^{2\pi} \mathrm{d}\phi \int\limits_{-\infty}^{\infty} \mathrm{d}x' \, p(x', \phi) h[r \cos(\phi - \varphi) - x'] \quad (3.14)$$

and let's replace the Radon coordinates x', ϕ with the coordinates of the fan beam sinogram by using the inverse relations of Equation 3.13:

$$x'(u) = \frac{D_1 u}{\sqrt{D_1^2 + u^2}} ,$$

$$\phi(u, \beta) = \beta - \tan^{-1} \frac{u}{D_1} , \quad (3.15)$$

for the case of flat detector and

$$x'(\gamma) = -D_1 \sin\gamma ,$$

$$\phi(\gamma, \beta) = \beta + \gamma, \quad (3.16)$$

for the case of curved detector. By changing into Equation 3.14 the variables defined in Equation 3.15 for the flat detector and Equation 3.16 for the curved detector, and by using the following auxiliary functions (see Figure 3.10)

$$U(r, \varphi, \beta) = D_1 + r \sin(\varphi - \beta) ,$$

$$L(r, \varphi, \beta) = \sqrt{U^2(r, \varphi, \beta) + r^2 \cos^2(\varphi - \beta)} \quad (3.17)$$

we obtain the following inversion formulae for the fan beam data (see Kak and Slaney [14] for the complete derivation):

$$f(r, \varphi) = \frac{1}{2} \int\limits_0^{2\pi} \frac{1}{U^2} \mathrm{d}\beta \int\limits_{-\infty}^{\infty} \mathrm{d}u \left(\frac{D_1}{\sqrt{D_1^2 + u^2}} \right) g^{(f)}(u, \beta) h(u' - u)$$

$$(3.18)$$

$$= \frac{1}{2} \int\limits_0^{2\pi} \frac{1}{L^2} \mathrm{d}\beta \int\limits_{-\infty}^{\infty} \mathrm{d}\gamma \, (D_1 \cos\gamma) g^{(c)}(\gamma, \beta) h^{(c)}(\gamma' - \gamma)$$

$$(3.19)$$

where $h^{(c)}$ is a weighted version of the ramp filter given by

$$h^{(c)}(\gamma) = \left(\frac{\gamma}{\sin\gamma}\right)^2 h(\gamma) . \qquad (3.20)$$

In order to perform the backprojection step properly, it is useful to write the equations relating the image pixel (cylindrical) coordinates (r, φ) to its projection point on the detector $(u(r, \varphi, \beta)$ or $\gamma(r, \varphi, \beta)$ depending on the shape of the detector) for a given gantry angle β:

$$u(r, \varphi, \beta) = \frac{D}{U(r, \varphi, \beta)} r \cos(\varphi - \beta) ,$$

$$\gamma(r, \varphi, \beta) = \begin{cases} \arccos \frac{U(r,\varphi,\beta)}{L(r,\varphi,\beta)} & \text{if } (\varphi - \beta) \in [0; \frac{\pi}{2}[\cup [\frac{3\pi}{2}; 2\pi[\\ -\arccos \frac{U(r,\varphi,\beta)}{L(r,\varphi,\beta)} & \text{if } (\varphi - \beta) \in [\frac{\pi}{2}; \frac{3\pi}{2}[\end{cases} .$$

$$(3.21)$$

The above inversion formulae can be used for reconstruction of fan beam data from flat (Equation 3.18) and curved detectors (Equation 3.19). As one can see, they are in the form of filtered backprojection, with some difference with respect to the parallel beam formula. First, the fan beam data must be pre-weighted with a factor that only depends on the radial coordinate $(D_1/\sqrt{D_1^2 + u^2}$ for the flat detector and $D_1 \cos\gamma$ for the curved one). The filter kernel is the standard ramp filter, $\tilde{h}$ (see Equations 3.6 and 3.8), for the case of flat detector. For the curved detector, the coordinate transformation (parallel to fan) introduced a bit different, weighted form for the filter kernel.

The backprojection is done by taking into account a space-dependent weighting factor (U and L for the flat and curved detectors, respectively) as defined in Equation 3.17. The geometrical interpretation of these factors can be understood by looking at Figure 3.10. For the purpose of keeping the next equations as compact as possible we also define the four weighting factors:

$$W_{\mathrm{BP}}^{(f)}(r, \varphi, \beta) = U^2(r, \varphi, \beta) \; ; \; w^{(f)}(u) = \frac{D_1}{\sqrt{D_1^2 + u^2}} \; ,$$

$$W_{\mathrm{BP}}^{(c)}(r, \varphi, \beta) = L^2(r, \varphi, \beta) \; ; \; w^{(c)}(\gamma) = D_1 \cos \gamma \; .$$

$$(3.22)$$

where the $w^{(\cdot)}$ factors are applied to the fan beam data before filtering, and the $W_{\mathrm{BP}}^{(\cdot)}$ factors are applied on backprojection.

The reader should note that the integration limit in the backprojection integral has been changed from π to 2π and a factor $1/2$ has been added to compensate for the effect of this change. The reconstruction formulae in Equations 3.18 and 3.19 must be used assuming that the fan beam data was acquired on a rotation interval of 2π (i.e., 360 degrees). This acquisition modality is referred to as *full scan* acquisition. Similarly to the parallel beam geometry, also in this case the ramp filter is commonly modified in practice by mean of apodisation windows, and then truncated in the frequency space according to the sampling theory (see Equation 3.5).

We can now write the reconstruction formulae for the FBP algorithm in fan beam geometry:

$$\tilde{f}_{\mathrm{FBP}}^{(\cdot)}(r, \varphi) = \frac{1}{2} \int_0^{2\pi} \frac{\mathrm{d}\beta}{W_{\mathrm{BP}}^{(\cdot)}} \, \tilde{q}^{(\cdot)} \; ,$$

$$(3.23)$$

$$\tilde{q}^{(f)}(u, \beta) = \left[(w^{(f)} g^{(f)}) * \tilde{h} \right](u, \beta) = (g_w^{(f)} * \tilde{h})(u, \beta) \; ,$$

$$(3.24)$$

$$\tilde{q}^{(c)}(\gamma, \beta) = \left[(w^{(c)} g^{(c)}) * \tilde{h}^{(c)} \right](\gamma, \beta) = (g_w^{(c)} * \tilde{h}^{(c)})(\gamma, \beta) \; .$$

$$(3.25)$$

where $\tilde{q}^{(\cdot)}$ are the filtered fan beam sinograms (in a given detector configuration), $\tilde{h}$ is the apodised ramp filter defined in Equation 3.5, and $\widetilde{h}^{(c)}$ is a modified version of the curved detector ramp filter (see Equation 3.20) modified to take into account apodisation and truncation, i.e., $\widetilde{h}^{(c)}(\gamma) = (\gamma/\sin\gamma)^2 \widetilde{h}(\gamma)$. We have also used the notation $g_w^{(\cdot)}$ for the weighted fan beam sinogram. A possible recipe for the implementation of 2D reconstruction of fan beam data using FBP is reported in Algorithm 3.

Algorithm 3 - FBP, 2D FAN BEAM, FULL-SCAN

---

1. Select the apodisation window $A(\nu)$ and compute the discrete version of the modified filter kernel $\widetilde{h}$;

2. if curved detector geometry is employed, multiply the modified filter kernel by the factor $(\gamma/\sin\gamma)^2$;

3. multiply the fan beam sinogram $g^{(\cdot)}$ by the radial weighting factor $w^{(\cdot)}$, obtaining the weighted fan beam sinogram $g_w^{(\cdot)}$;

4. apply the 1D ramp filter to the weighted fan beam sinogram: for each available projection angle β, take the 1D DFT's $G_w^{(\cdot)} = \mathcal{F}_1 g_w^{(\cdot)}$ and $\widetilde{H} = \mathcal{F}_1 \widetilde{h}$ of the weighted fan beam sinogram and of the modified filter, respectively, and multiply them in the frequency domain; afterwards, compute the filtered sinogram $\widetilde{q}^{(\cdot)}$ as the inverse 1D DFT of the product $G_w^{(\cdot)} \cdot H$, i.e., $\widetilde{q}^{(\cdot)} = \mathcal{F}_1^{-1}(G_w^{(\cdot)} \cdot \widetilde{H})$;

5. reconstruct the image by backprojecting each row of the weighted filtered fan beam sinogram $\widetilde{q}^{(\cdot)}/W_{\mathrm{BP}}^{(\cdot)}$ on the image plane; the backprojection is performed by following the original direction of each acquired line integral at each gantry angle;

6. multiply the entire reconstructed image by $1/2$.

3.2.3 Python implementation of the fan beam FBP

The following implementation is based on the Jupyter notebook called FBPFanBeamDemo, included in the companion software (see Chapter 2 for instructions). In this example, we will use a voxelised version of the Shepp-Logan 2D phantom and we will generate its fan beam sinogram using the SinogramGenerator class of the DAPHNE framework.

In the geometric setup, we will use a planar detector for this example (as it is more usual for micro-CT users, like the authors of this book). The following imports are required for the selected example:

```
1 from Algorithms.SinogramGenerator import
      SinogramGenerator
2 from Algorithms.FBP import FBP
3 from Geometry.ExperimentalSetupCT import
      ExperimentalSetupCT,Mode,DetectorShape
4 from Misc.Preview import Visualize3dImage
```

Listing 3.18: Modules required to perform FBP fan beam reconstruction.

After setting up the system geometry (see the FBPFanBeamDemo notebook for the details, or follow the instructions on Chapter 2 to create custom geometries), we can run the following three lines of code to load the binary data of the Shepp-Logan phantom and to generate its fan beam sinogram:

```
1 # Read the phantom data into a numpy array
2 phantom = np.fromfile("../Data/SL_HC_256x256.raw",dtype=
      np.float32).reshape((256,256,1))
3
4 # Create an instance of the sinogram based on the
      experimental setup
5 s=SinogramGenerator(my_experimental_setup)
6
7 # Generate the sinogram of the phantom using a Siddon
      algorithm
8 sino=s.GenerateObjectSinogram(np.flipud(phantom),
      transpose_image=1)
```

Listing 3.19: Code required to load the phantom data and generate its fan beam sinogram.

When calling the GenerateObjectSinogram function, we have first flipped the phantom data vertically using the *Numpy* function flipud(), in order to give negative y values to the first rows of phantom stored in memory. Moreover, we have enabled the transpose_image flag to take into account the internal storage order of the arrays in *Numpy*. The resulting sinogram is shown in Figure 3.11.

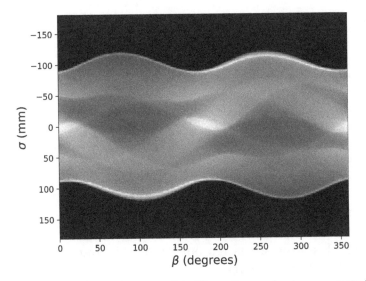

Figure 3.11: Fan beam sinogram of the Shepp-Logan phantom, generated using the Siddon algorithm.

The next step is to define some auxiliary variables and data structures that will be useful in the subsequent implementation of the algorithm. These variables are initialised in the listing 3.20.

```
1  # Set the side of a squared target image
2  Imsize = 256
3  N_rad = my_experimental_setup.
       _number_of_pixels_per_slice
4  N_ang = my_experimental_setup._number_of_gantry_angles
5
6  # Create an array with the radial coordinates
7  # of the detector pixels
8  D = my_experimental_setup._sad_mm
9  mag = my_experimental_setup._sdd_mm /
       my_experimental_setup._sad_mm
10 virt_det_pitch_mm = my_experimental_setup.
       _detector_pitch_mm / mag
11 u_det = (np.arange(N_rad) - N_rad // 2 + 0.5) *
       virt_det_pitch_mm
12
13 # Grid of the coordinates of the pixel centers
14 # in the target image (pitch=1 for simplicity)
15 # 1 - Cartesian coordinates
```

```
16 x,y = np.mgrid[:Imsize, :Imsize] - Imsize // 2 + 0.5
17 y = y*(-1) # This line allow to set the y axis
       increasing upwards
18 # 2 - Polar coordinates
19 r=np.sqrt(np.square(x)+np.square(y))
20 varphi=np.arctan2(y,x)
21
22 # Create an array with the gantry angles
23 beta_deg = np.arange(0, N_ang, 1)
```

Listing 3.20: Auxiliary variables and data structures used in the implementation of the fan beam FBP reconstruction.

We are now ready to apply the radial weighting function and subsequent ramp filtering to the fan beam sinogram. Before that, we will just create an instance f of the FBP class, and we assign our Sinogram object (as generated by the GenerateObjectSinogram() function).

```
1  # Set up the FBP algorithm
2  f=FBP()
3  f.sinogram=sino
4  f.interpolator='cubic'
5
6  # Sinogram weighting factor
7  w = D / np.sqrt(np.square(D) + np.square(u_det))
8
9  # Weight sinogram
10 f.sinogram.data = f.sinogram.data * w
11
12 # Apply ramp filter
13 # (flat detector, same kernel of the parallel beam case)
14 f.FilterSino()
```

Listing 3.21: Instantiation of a new **FBP** object and assignment of the previously generated **Sinogram** object. After assignment to **f**, the sinogram is weighted by the proper function before ramp filtering.

Even though DAPHNE provides high-level implementation of the FBP algorithm with flexible choice of geometry setup, we decided to avoid such a shortcut and expose all the relevant steps in the listings. This should be more instructive for the reader to understand the implementation details. The last line of listing 3.21 calls the GenerateRamp() function shown in listing 3.8 on page 40, and then applies the Ram-Lak filter to the sinogram in the frequency domain as already done for the parallel beam example.

Following the example for the parallel beam geometry, we will first show how to create a partial reconstruction by backprojection of one single filtered projection (this time at gantry angle $\beta = 0$). The main change with respect to the parallel beam backprojection (listing 3.13 on page 47 is in the definition of the detector coordinate u at which the ray starting from the source and passing through the (r, φ) point of the image plane (see Equations 5.5). Moreover, the values accumulated at the destination must be properly weighted as defined in Equations 3.22.

```
1  # This tuple stores the projection (1D array)
2  # at the selected angle and the corresponding
3  # gantry angle in radians (0 in this example)
4  id_proj = 0
5  (proj, beta_rad) = (s_filt[:,id_proj],
6                      np.deg2rad(beta_deg)[id_proj])
7
8  # Variables used in backprojection
9  U = D + r * np.sin(varphi-beta_rad)
10 u = r * np.cos(varphi-beta_rad) * D / U
11 W_BP = np.square(U)
12
13 # Destination image
14 img_partial = np.zeros([Imsize,Imsize], dtype=float)
15
16 # Backproject
17 img_partial += interp1d(
18     x = u_det,
19     y = proj,
20     kind = 'linear',
21     bounds_error = False,
22     fill_value = 0,
23     assume_sorted = False
24 ) (u) / W_BP
```

Listing 3.22: Backprojection of a single filtered projection at $\beta = 0$, in fan beam geometry (flat detector).

Figure 3.12 shows the result of the fan beam full scan (filtered) backprojection, along with some limited-arc reconstructions.

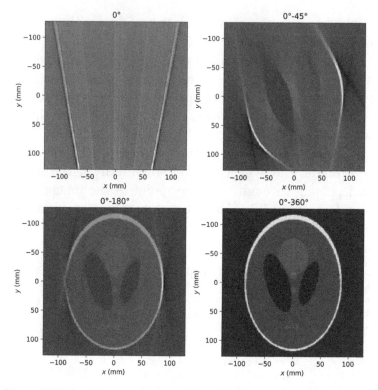

Figure 3.12: Partial reconstructions (top-left, top-right and bottom-left) and full reconstruction (bottom-right) of the Shepp-Logan phantom in fan beam geometry.

3.2.4 Data redundancy and short-scan reconstruction

The last step in the previous algorithm for full-scan fan beam FBP consists of dividing the reconstructed image by 2, as needed by directly applying Equation 3.23. This normalization step is necessary because the fan beam data acquired on a 2π interval is redundant, i.e., every line integral x', ϕ crossing the object is acquired exactly twice. More specifically, the periodicity of the fan beam sinogram is:

$$g^{(f)}(u, \beta) = g^{(f)}(-u, \beta + \pi - 2\tan^{-1}\frac{u}{D_1})$$
$$g^{(c)}(\gamma, \beta) = g^{(c)}(-\gamma, \beta + \pi - 2\gamma). \tag{3.26}$$

Hence, when a full scan is performed, each line integral acquired at a given gantry angle β has a redundant one (acquired in the opposite direction) in the projection taken at $\beta' = \beta + \pi - 2\gamma$ (this notation also applies to a flat detector, where $\gamma = \tan^{-1} \frac{u}{D_1}$)). We will now derive the minimum rotation interval of the gantry that allows us to obtain a minimally redundant complete dataset, i.e., a fan beam dataset with the smallest number of redundant line integrals, but with every line integral acquired at least once. First of all, let us notice from Equation 3.13 that

$$\beta = \phi + \gamma = \phi - \sin^{-1} \frac{x'}{D_1} , \qquad (3.27)$$

and then, for a given angle ϕ of the Radon space, the radial sampling inside a circle of radius x'_{FOV} is completed by varying the gantry angle β in the range

$$\phi - \sin^{-1} \frac{x'_{\text{FOV}}}{D_1} \leqslant \beta < \phi + \sin^{-1} \frac{x'_{\text{FOV}}}{D_1} , \qquad (3.28)$$

or, equivalently,

$$\phi - \gamma'_{\text{FOV}} \leqslant \beta < \phi + \gamma'_{\text{FOV}} , \qquad (3.29)$$

where γ'_{FOV} is the semi-aperture of the detected X-ray fan beam. A complete dataset, in the sense of Radon inversion, is such that all the angles $\phi \in [0, \pi[$ must be acquired; thus, Equation 3.29 can be rewritten as

$$-\gamma'_{\text{FOV}} \leqslant \beta < \pi + \gamma'_{\text{FOV}}. \qquad (3.30)$$

The choice of the initial gantry angle is arbitrary, so the previous equation can be rewritten as:

$$0 \leqslant \beta < \pi + 2\gamma'_{\text{FOV}} . \qquad (3.31)$$

The minimum rotation interval leading to a complete fan beam dataset is then $\beta \in [0, \pi + 2\gamma'_{\text{FOV}}]$, i.e, a half-rotation plus the angular aperture of the detected X-ray fan beam.

A minimally redundant acquisition in CT is also referred to as a *short scan* acquisition. The advantages of short-scan acquisitions in CT over full-scan acquisitions are mainly related to the improvement of temporal resolution, as the time required for gathering a complete dataset is almost halved with respect to the gantry revolution time. From the reconstruction point of view, a short-scan dataset can't be reconstructed by simple application of the formulae in Equation 3.23 because it contains partially redundant data. In other words, some line integrals are acquired once while some others are acquired twice. In this case, the application of a global normalization factor (as we have done by the $1/2$ factor in Equation 3.23) just won't work.

One possible solution to the problem of partial redundancy is to apply a weighting window such that $w_{\mathrm{Short}} = 1/2$ on the redundant area and $w_{\mathrm{Short}} = 1$ elsewhere. This solution would cause a discontinuity along the radial direction in the sinogram, leading to severe artifacts when the ramp filter is applied. Parker [24] proposed a smooth weighting window with continuous derivatives:

$$
w_{\mathrm{Short}}(\gamma,\beta) = \begin{cases} \sin^2\left(\dfrac{\pi}{4}\dfrac{\beta}{\gamma_{\mathrm{FOV}}+\gamma}\right) & \text{if } 0 \leqslant \beta < 2(\gamma+\gamma_{\mathrm{FOV}}) \\ 1 & \text{if } 2(\gamma+\gamma_{\mathrm{FOV}}) \leqslant \beta < \gamma+\pi \\ \sin^2\left(\dfrac{\pi}{4}\dfrac{\pi+2\gamma_{\mathrm{FOV}}-\beta}{\gamma_{\mathrm{FOV}}-\gamma}\right) & \text{if } \gamma+\pi \leqslant \beta < \pi+2\gamma_{\mathrm{FOV}} \\ 0 & \text{if } \pi+2\gamma_{\mathrm{FOV}} \leqslant \beta < 2\pi \end{cases}
$$

$$(3.32)$$

The short-scan weighting window above has been written using curved detector coordinates, but it can be easily adapted for the flat detector by using $\gamma = \tan^{-1}(u/D_1)$. Reconstruction of short-scan fan beam data requires then that the fan beam sinogram g in Equations 3.24 and 3.25 must be further pre-weighted by a short-scan weighting window (e.g., like the Parker window in Equation 3.32) in order to pre-correct for partial data redundancy. For short scan reconstruction, we can now write

$$\widetilde{f}^{(\cdot)}_{\text{FBP,Short}} = \int_0^{\pi+2\gamma_{\text{FOV}}} \frac{\mathrm{d}\beta}{W^{(\cdot)}_{\text{BP}}} \, \widetilde{q}^{(\cdot)}_{\text{Short}} \,, \tag{3.33}$$

where

$$\widetilde{q}^{(\cdot)}_{\text{Short}} = \left[\left(w^{(\cdot)} g^{(\cdot)} w_{\text{Short}} \right) \right] * \widetilde{h}^{(\cdot)} \,. \tag{3.34}$$

For completeness, we report here the Algorithm 4 for short-scan fan beam FBP reconstruction even though the only additional step required as compared to the full-scan algorithm is the pre-weighting with the short-scan window w_{Short} which replaces the global weighting by the 1/2 factor:

Algorithm 4 - FBP, 2D FAN BEAM, SHORT-SCAN

1. Multiply the short-scan sinogram by the short-scan weighting window w_{Short}.

2. Execute the same steps of the Algorithm 3 reported above (page 57), except for the last step of global weighting by the factor 1/2.

3.3 RECONSTRUCTION OF FAN BEAM DATA FROM HELICAL SCANS

Helical scans (or spiral scans) are special types of CT acquisition protocols in which the rotation of the gantry is made simultaneously with the linear translation of the patient bed. In the context of image reconstruction, it is customary to assume that the object function is still in the laboratory coordinate system while the source performs an helical trajectory. Let us first remember that the source trajectory in fan beam 2D geometry is described by the vector $\mathbf{x}_{\text{foc}}$:

$$\mathbf{x}_{\text{foc}}(\beta) = \begin{pmatrix} D_1 \sin \beta \\ -D_1 \cos \beta \\ z_{\text{start}} \end{pmatrix} \qquad (3.35)$$

where z_{start} is the axial position of the plane defining the circular trajectory. For standard helical scan (i.e., with constant radius and pitch), the source trajectory is instead

$$\mathbf{x}_{\text{foc}}(\beta) = \begin{pmatrix} D_1 \sin \beta \\ -D_1 \cos \beta \\ z_{\text{start}} + \dfrac{d \cdot \beta}{2\pi} \end{pmatrix},$$

$$\beta \in [0; \beta_{\text{stop}}],$$

$$z_{\text{foc}} \in \left[z_{\text{start}}; z_{\text{stop}} = z_{\text{start}} + \dfrac{d \cdot \beta_{\text{stop}}}{2\pi} \right], \qquad (3.36)$$

with d being the helical pitch, and z_{start} the starting z-position of the source.

Practical helical scan reconstruction is done by interpolation methods. In other words, we won't search for new inversion formulae for the fan beam sinogram but instead we attempt to create a synthetic (i.e., interpolated) version of the standard planar 2D fan beam sinogram by using the available projection data. Let us denote by z^* an arbitrary axial position at which we want to reconstruct a 2D slice, with $z_{\text{start}} < z^* < z_{\text{stop}}$. The simplest algorithm is obtained by linear interpolation of the required line integrals $g^{(\cdot)}(\cdot, \beta)$ from the projection data of two consecutive gantry rotations, let's say j and $j+1$, with $j = 0, \ldots, N-1$ where N is the number of gantry revolutions, and where j must be such that

$$z_{\text{start}} + jd < z^* < z_{\text{start}} + (j+1)d. \qquad (3.37)$$

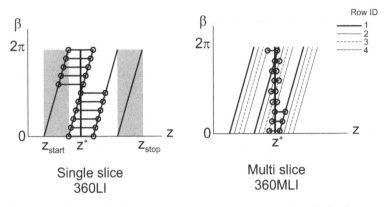

Figure 3.13: (Left) Rebinning scheme for a single-slice helical scan with the method 360LI. (Right) When the number of detector rows is increased, the z-distance of two useful data points for interpolation is reduced but is not constant throughout all the gantry angles (this figure refers to a 4-slice scanner). The depicted rebinning method for MSCT is called 360MLI.

For full-scan reconstruction, we can see from the first $\beta(z)$ graph of Figure 3.13 that the required angular range of the gantry to build a complete fan beam sinogram is 4π; moreover, the axial positions $z_{\text{start}} < z^* < z_{\text{start}} + d$ are unaccessible, as well as the positions $z_{\text{stop}} - d < z^* < z_{\text{stop}}$. This means that two segments of length d at the beginning and at the end of the helical trajectory cannot be reconstructed, even though they have been crossed by the X-ray fan beam for at least one gantry angle. Following the notation of Kalender [15] and Schaller [27], we denote the reconstruction method just described as $360°\text{LI}$, where LI stands for linear interpolation.

For MDCT scanners, the reader must figure out that each detector row defines a $\beta(z)$ graph shifted by a factor $k\Delta z$ with respect to the central row, where Δz is the axial extent of a given row and k is the row index, with $k = -M/2, \ldots, M/2 - 1$ and M is the number of rows. Optimal reconstruction is done for each arbitrary slice position z^* by searching the two closest data points for each gantry angle β among all the available detector rows. The extension of the 360LI used method for helical SSCT is denoted by 360MLI in the case of helical MDCT.

In most cases, the periodicity condition of the fan beam sinogram is exploited to further reduce the z-distance of the data points to be interpolated. Using this method leads to a family of rebinning algorithms called 180LI (for SSCT) and 180MLI (for MDCT). The reader can refer to Shaller [27] for further reading on the rebinning techniques in MSCT.

3.4   3D FBP IN CONE BEAM GEOMETRY

The inversion of the 3D Radon transform in $\mathbb{R}^3$ is obtained by the following equation (see Natterer [21]):

$$f(\mathbf{x}) = -\frac{1}{8\pi^2} \mathcal{R}_3^{\#} \left[\frac{\partial^2}{\partial r^2} p(r, \boldsymbol{\alpha}) \right]. \qquad (3.38)$$

The application of the above equation to image reconstruction from real data has been the subject of intensive study by several investigators for decades. From a practical point of view, we are interested in the reconstruction of projection data acquired in *cone beam* geometry, as this is the geometry employed in volumetric CT scanners using 2D (curved or flat) detectors.

As a first practical problem, data completeness in the Radon domain is required in order to apply Equation 3.38. Tuy and Smith have shown that an exact reconstruction is possible only if all the planes crossing the object do intersect the source trajectory in at least one point. This is also called the *Tuy-Smith data sufficiency condition* (DSC) [32]. Based on Tuy's inversion formula, Grangeat (Grangeat 1990 [10]) proposed a reconstruction framework in which the radial derivative of the 3D Radon transform can be extracted from line integrals acquired in cone beam geometry. Nevertheless, this solution presents practical issues and numerical instabilities.

Exact 3D CT reconstruction has limited application in real-world CBCT imaging. The most used source trajectory for most CT and micro-CT scanners is a circle, which does not fulfill the Tuy-Smith sufficiency condition. Several scanning trajectories have been evaluated by investigators in order to handle the data incompleteness, such as perpendicular circles or circles plus lines (Kudo and Saito 1994 [18]), a circle plus arc (Wang 1999 [37]), or a helix with constant or variable pitch and radius (Katsevich 2004 [16]). Among them, only the circle and helix with constant pitch and radius have found significant applications. Even though the helical cone beam scan is performed on modern multi-slice CT (MSCT) scanners the computational burden of exact algorithms is still an issue. Hence, approximate algorithms are by far most used rather than exact ones for practical cone beam reconstruction. The most widespread method for approximate cone beam reconstruction was derived by Feldkamp, Davis and Kress (Feldkamp 1984 [6]) and is described in this section.

3.4.1 The Feldkamp-Davis-Kress (FDK) method

The circular cone beam geometry is depicted in Figure 3.14. The data point coordinates are very similar to those of fan-beam geometry, where a longitudinal (i.e., axial) coordinate v was added taking into account the axial extension of the 2D detector. The basic idea behind the Feldkamp method, or FDK method, is that for moderate axial apertures of the detected cone beam, the acquisition geometry should not deviate too much from a multi-slice fan beam acquisition. The discrepancy between cone beam and multi fan beam geometry could then be compensated by means of a correction factor. Furthermore, the resulting algorithm should reduce to the fan beam algorithm when just the axial midplane is reconstructed.

Let us first consider the case of the flat detector, similarly as Feldkamp and colleagues have done in their

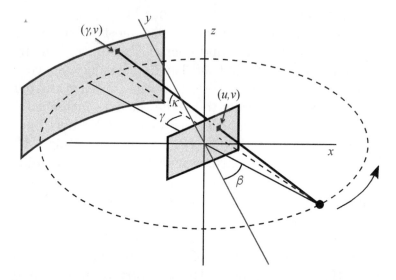

Figure 3.14: Meaning of the coordinates in circular cone beam geometry. The axial coordinate has been denoted by v for both the flat and curved detectors, due to the fact that they have identical geometrical meaning (apart from a scaling factor).

original article. The cone beam coordinates are denoted by u, v, β, where v is the axial coordinate of the detector and the other two variables have the same meaning as those used for the fan beam geometry. In analogy with the notation used for fan beam sinograms, we will denote the cone beam projections as $g^{(f)}(u, v, \beta)$. As one can see from Figure 3.14, each detector row ($v = \mathrm{cost}$) defines a tilted fan beam with tilt angle κ. Let us first rewrite the fan beam reconstruction formula, which should provide exact reconstruction of the object at $z = 0$. For the flat detector,

$$
\widetilde{f}_{\mathrm{FBP}}(x, y, 0) = \widetilde{f}_{\mathrm{FDK}}(\mathbf{x})\rvert_{z=0}
$$

$$
= \frac{1}{2} \int\limits_{0}^{2\pi} \frac{1}{U^2} \mathrm{d}\beta \int\limits_{-\infty}^{\infty} \mathrm{d}u \left(\frac{D_1}{\sqrt{D_1^2 + u^2}} \right) g^{(f)}(u, 0, \beta) h(u' - u).
$$

$$
(3.39)
$$

Away from the midplane, we can see that the line integrals in the tilted fan beam are scaled by a factor

$$\sqrt{D_1^2 + u^2 + v^2}/\sqrt{D_1^2 + u^2} \tag{3.40}$$

with respect to those passing from the central row of the detector. Hence, an approximate reconstruction for $z \neq 0$ can be obtained by compensating for this extra-length of the tilted line integrals in Equation 3.39:

$$
\begin{aligned}
\widetilde{f}_{\text{FDK}}(\mathbf{x}) &= \frac{1}{2} \int_0^{2\pi} \frac{1}{U^2} d\beta \int_{-\infty}^{\infty} du \left(\frac{D_1}{\sqrt{D_1^2 + u^2}} \frac{\sqrt{D_1^2 + u^2}}{\sqrt{D_1^2 + u^2 + v^2}} \right) \\
&\quad \cdot g^{(f)}(u, v, \beta) h(u' - u) \\
&= \frac{1}{2} \int_0^{2\pi} \frac{1}{U^2} d\beta \int_{-\infty}^{\infty} du \left(\frac{D_1}{\sqrt{D_1^2 + u^2 + v^2}} \right) g^{(f)}(u, v, \beta) h(u' - u).
\end{aligned}
\tag{3.41}
$$

For the curved detector, the scaling factor in the tilted fans is simply $\sqrt{D_1^2 + v^2}/D_1$ so that the FDK reconstruction formula can be written as

$$
\begin{aligned}
\widetilde{f}_{\text{FDK}}(\mathbf{x}) &= \frac{1}{2} \int_0^{2\pi} \frac{1}{L^2} d\beta \int_{-\infty}^{\infty} d\gamma \left(\frac{D_1^2 \cos \gamma}{\sqrt{D_1^2 + v^2}} \right) \\
&\quad \cdot g^{(c)}(\gamma, v, \beta) h^{(c)}(\gamma' - \gamma).
\end{aligned}
\tag{3.42}
$$

It is worth noting that the ramp filtering in the FDK method is done row-by-row similarly to the standard fan beam reconstruction. As one can see from Equations 3.41 and 3.42, FDK differs from fan beam FBP just for the different shape of the pre-weighting factor, $w^{(\cdot)}$, that is also dependent on the axial position v:

$$
\begin{aligned}
w_{\text{FDK}}^{(f)}(u, v) &= \frac{D_1}{\sqrt{D_1^2 + u^2 + v^2}}, \\
w_{\text{FDK}}^{(c)}(\gamma, v) &= \frac{D_1^2 \cos \gamma}{\sqrt{D_1^2 + v^2}},
\end{aligned}
\tag{3.43}
$$

while the backprojection weighting factors $W_{\mathrm{BP}}^{(\cdot)}$ and the ramp filter $\tilde{h}$ are kept identical to those of the fan beam formula. In this case, the backprojection is done in the 3D space by following the original direction of the acquired line integrals. Even if the basic steps are very similar to those reported above for the full-scan fan beam FBP, we summarize here all the steps required for FDK reconstruction:

Algorithm 5 - FBP, 3D CONE BEAM (FDK), FULL-SCAN

---

1. Select the apodisation window $A(\nu)$ and compute the discrete version of the modified filter kernel $\tilde{h}$;

2. if curved detector geometry is employed, multiply the modified filter kernel by the factor $(\gamma/\sin\gamma)^2$;

3. multiply the cone beam projection $g^{(\cdot)}$ by the weighting factor $w_{\mathrm{FDK}}^{(\cdot)}$ (see Equation 3.43), obtaining the weighted CB projection $g_w^{(\cdot)}$;

4. apply the 1D ramp filter to each row of the weighted CB projections: for each available projection angle β and for each axial position v of the detector, take the 1D DFT's $G_w^{(\cdot)} = \mathcal{F}_1 g_w^{(\cdot)}$ and $\tilde{H} = \mathcal{F}_1 \tilde{h}$ of the weighted CB projection and of the modified filter, respectively, and multiply them in the frequency domain; afterwards, compute each row of the filtered CB projections $\tilde{q}^{(\cdot)}$ as the inverse 1D DFT of the product $G_w^{(\cdot)} \cdot H$, i.e., $\tilde{q}^{(\cdot)} = \mathcal{F}_1^{-1}(G_w^{(\cdot)} \cdot \tilde{H})$;

5. reconstruct the image by backprojecting each weighted filtered CB projection $\tilde{q}^{(\cdot)}/W_{\mathrm{BP}}^{(\cdot)}$ on the 3D image space; the backprojection is performed by following the original direction of each acquired line integral at each gantry angle.

6. Multiply the entire reconstructed image by $1/2$.

Short-scan reconstruction of CB data by the FDK method is also possible, even though the data redundancy away from the midplane is not guaranteed for all object functions and hence direct row-wise application of the short-scan weighting window for fan beam geometry can lead to image artifacts (see, for instance, Maaß et al.

2010 [19]). Interestingly, as reported in the appendix of their original article, Feldkamp *et al.* [6] have shown that their method is exact for z-invariant object functions, i.e., when f is such that

$$\frac{\partial f(x, y, z)}{\partial z} = 0. \tag{3.44}$$

In this case, it is easy to see that the line integrals away from the midplane are just the scaled version of the line integrals in the $z = 0$ plane, i.e.:

$$
\begin{aligned}
g^{(f)}(u, v, \beta) &= \frac{g^{(f)}(u, 0, \beta)}{w_{\mathrm{FDK}}^{(f)}(u, v)}, \\
g^{(c)}(\gamma, v, \beta) &= \frac{g^{(c)}(\gamma, 0, \beta)}{w_{\mathrm{FDK}}^{(c)}(\gamma, v)}.
\end{aligned}
\tag{3.45}
$$

For this class of objects, data redundancy and periodicity conditions for the CB projections can be found similarly to the fan beam case. This property can have practical applications: for instance, Panetta *et al.* [23] have exploited the redundancy of CB projections of z-invariant objects for the determination of a subset of the 7 misalignment parameters of CBCT systems.

In practice, the basic idea behind the FDK approximation is the following: tilted X-ray paths can be approximated by rays that are parallel to the xy plane, and the error is compensated for by a correction (scaling) factor. This approximation becomes exact in the case of globally or even locally z-invariant objects, which is realistic when the cone angle κ is small. For "normal" objects, varying weakly along z, the approximated nature of the FDK method is visible as underestimations of the reconstructed values for increasing cone angles. The difficulty in keeping the image quality of peripheral slices similar to central slices can be viewed as one of the reasons for the end of the so-called "slice war". Several decades after the publication of the article of Feldkamp *et al.*, there is still an effort to modify and extend the FDK reconstruction formula. In 1993, Wang *et al.* generalized the FDK method to an arbitrary source trajectory [36]. In 2002, the same author proposed a different formulation that allows us to reconstruct data using a transversally shifted detector array [35]; the big advantage of such a formulation is that one can virtually enlarge the scanner FOV and thus cope with the problem of a large object and small flat panel detectors.

3.4.2 Python implementation of the FDK algorithm

We will now show how to modify the programming example given for fan beam geometry, in order to perform fully 3D reconstruction with the FDK formula 3.41. Again, even though such a type of reconstruction is implemented at a high level (i.e., by calling a minimal number of task-specific functions) in DAPHNE , we will show here a more detailed, low-level implementation. In the following example, we will use the 3D version of the Shepp-Logan head phantom (see Kak and Slaney [14] for its definition), as shown in Figure 3.15. The full example is available in the Jupyter notebook called FDKDemo.ipynb of the companion software. Let us skip here the steps of loading and displaying the phantom data, as they are very similar to the procedure already employed for the preceding examples. It what follows, we will assume we have already loaded the phantom projection data into the projs array, with shape projs.shape=(256,256,360), where the first two dimensions are the horizontal (u) and vertical (v) sizes of the detector, and the last one is the number of projections over 360°.

The first step in the implementaion of the FDK algorithm is to generate a 2D weighting window w_{FDK} to be applied to the projection data. We will use a flat detector

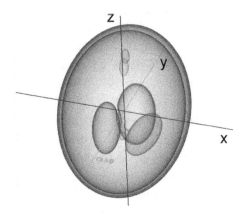

Figure 3.15: 3D version of the Shepp-Logan head phantom.

in this example, so we must generate the weighting window according to the first of the two Equations 3.43. The following listing allows us to perform this step.

```
1 # Define some geometry-related variables
2 D = 500
3 ProjSizeU = 256
4 ProjSizeV = 256
5 Nang = 360
6 AngStep = 1.0
7 VirtDetPitch = 1.421758727602353
8
9 # Create an array with the radial and axial coordinates
10 # of the 2D detector pixels
11 u_det,v_det = np.mgrid[-ProjSizeU // 2 + 0.5 : ProjSizeU
      // 2 + 0.5 : 1,
12                        -ProjSizeV // 2 + 0.5 : ProjSizeV
      // 2 + 0.5 : 1
13                        ] * VirtDetPitch
14
15
16 # Generate the weighting window (as a 3D array)
17 w = D / np.sqrt(np.square(D) + np.square(u_det) + np.
      square(v_det))
18 w = np.repeat(w[:, :, np.newaxis], Nang, axis=2)
19
20 # Apply weighting factors to all projections
21 weighted_projs = w*projs
```

Listing 3.23: Pre-filtering data weighting in the FDK algorithm.

The geometry variables shown in listing 3.23 are numerically identical to those employed in the fan beam example above (Section 3.2). They arise from a geometry setup in which $D_1 = 500$ mm, $D_2 = 250$ mm, $\Delta\beta = 1°$, and with a detector made up of 256 pixels along the radial direction, with an angular aperture of 40°. The *Numpy* function np.repeat() called at line 18 of the previous listing, created **Nang** identical copies of the weighting window. As a result, the pre-filter weighting of the entire dataset can be done by just multiplying the 3D array projs by w, as is done in line 21 of the listing 3.23.

The weighted projection data can be now ramp-filtered along the u (radial) direction. As explained above, in the case of a flat detector the shape of the ramp filter (both IR and FR) is identical to that of the parallel beam case. This can be easily obtained by calling the GenerateRamp() function. This time, the matrix **Hm** stores ProjSizeV*Nang copies of the 1D filter FR. Hence, the filtering step in the frequency domain can be carried out in a way formally similar to the parallel beam and fan beam cases:

```
# Ram-Lak filtering of the projection data
# Hm is a 3D array storing ProjSizeV*Nang copies
# of the 1D Ram-Lak filter frequency response
fft1d_projs = np.fft.fft(weighted_projs, axis=0)
filtered_projs = np.real(
        np.fft.ifft(fft1d_projs * rampFRmatrix, axis=0)
    )
```

Listing 3.24: Ramp filtering in the FDK algorithm.

Let us now declare the variables storing the geometrical parameter of the target image used for reconstruction, such as its sizes and the coordinates of each voxel (cartesian and cylindrical):

```
# Grid of the coordinates of the pixel centers
# in the target image (pitch=1 in this example)

# Num. of image voxels along x,y = Num. of detector
    pixels along radial dir.
```

```
 5 Imsize = ProjSizeU
 6 # This allows to control the number of slices (xy planes
       ) to be reconstructed
 7 # (impacts performance - select a small number of slices
       on non-performant PCs
 8 SliceStart = 0
 9 SliceEnd = ProjSizeV
10 Nslices = SliceEnd-SliceStart
11
12 # 3D Cartesian coordinate of the voxels in the target
       image
13 x,y,z = np.mgrid[:Imsize, :Imsize, SliceStart:SliceEnd]
       - Imsize // 2 + 0.5
14 y = y*(-1) # As done for fan-beam, this makes the y-axis
       to increase upwards
15
16 # 3D cylindrical coordinates, to be used in
       backprojection
17 r=np.sqrt(np.square(x)+np.square(y))
18 varphi=np.arctan2(y,x)
19 # z is the same in both coordinate types
```

Listing 3.25: Geometrical parameters of the target image and voxel coordinates.

The backprojection is performed in a way that is very similar to the fan beam case. In this case, we require bilinear interpolation to get the (filtered) projection data from the available points in the voxel driven backprojection. Performance reasons suggested that we use the *Scipy* function `RegularGridInterpolator` of the `scipy.interpolator` package. Figure 3.16 shows the result of FDK reconstruction for variable angular sampling step. In Figure 3.17, some common FDK artifacts are shown.

```
 1 from scipy.interpolate import RegularGridInterpolator
 2
 3 # These three variables allow controlling
 4 # the angular range and step in backprojection.
 5 ang_subsampling = 1
 6 first_ang_id = 0
 7 last_ang_id = Nang
 8
 9 # Destination image
10 img_fdk = np.zeros([Imsize,Imsize,Nslices], dtype=np.
       float32)
11
```

```
12  # Define the sets of filtered projections and
        corresponding gantry angles
13  fprojs_subset=np.transpose(
14      filtered_projs[:,:,first_ang_id:last_ang_id:
        ang_subsampling],
15      (2,0,1)
16  )
17  beta_deg = np.arange(0, Nang, 1)
18  angs=np.deg2rad(beta_deg)[first_ang_id:last_ang_id:
        ang_subsampling]
19
20  # Iterate over all the selected projections
21  for (fproj, beta_rad) in zip(fprojs_subset, angs):
22      print("Backprojecting, gantry angle (rad): ",
        beta_rad)
23      U = D + r * np.sin(varphi-beta_rad)
24      u = r * np.cos(varphi-beta_rad) * D / U
25      v = z * D / U
26      W_BP = np.square(U)
27
28      # Accumulate the weighted values in the destination
        image
29      img_fdk += RegularGridInterpolator(
30          points = (u_det_1d, v_det_1d),
31          values = fproj,
32          method = 'linear',
33          bounds_error = False,
34          fill_value = 0
35      ) ((u,v)) / W_BP
```

Listing 3.26: Projection filtering and cone beam backprojection in the FDK algorithm.

3.5 OTHER FOURIER-BASED METHODS

3.5.1 Backprojection-Filtration (BPF)

We have shown in Equation 3.11 that the backprojection operator $\mathcal{R}^{\#}$, which is the dual operator of the Radon transform $\mathcal{R}^{\#}$, is different from the inverse Radon transform $\mathcal{R}^{-1}$. When we omit the pedix in the RT symbol, it is understood that we are referring to the transform in two dimensions. We have seen in the previous section that we can still obtain an approximated (blurred) version of f by performing a simple, unfiltered backprojection of the sinogram, i.e., $\tilde{f}_{BP} = \mathcal{R}^{\#}p = (\mathcal{R}^{\#}\mathcal{R})f$. The function $\tilde{f}_{BP}$ is called the laminogram of the object

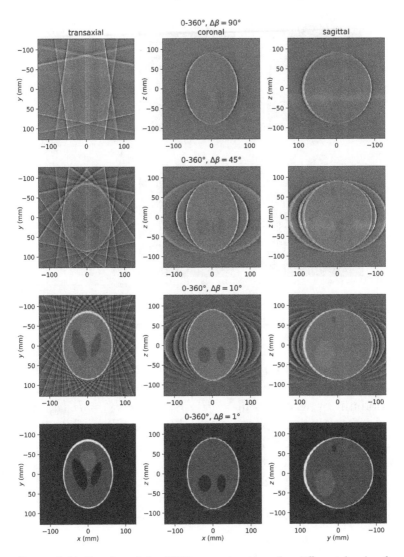

Figure 3.16: Results of the FDK reconstruction, for different levels of the `ang_subsampling` variable.

function f. We can state that the laminogram and the object function are linked by the operator $\mathcal{R}^{\#}\mathcal{R}$, having an impulse response h' such that

$$\widetilde{f}_{\mathrm{BP}} = \mathcal{R}^{\#}\mathcal{R}f = h' **f \tag{3.46}$$

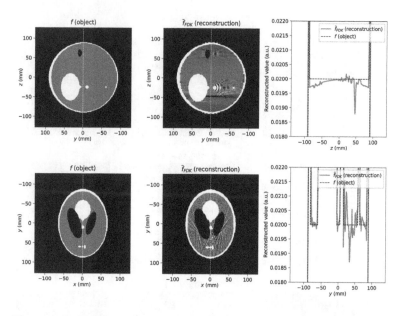

Figure 3.17: FDK image quality along z (top row) and y (bottom row).

In order to find an analytical expression for the 2D filter h', it is easier to study the problem in the frequency domain by writing $\widetilde{F}_{\text{BP}} = H' \cdot F$, where $H' = H'(u, v)$ is the frequency response of the blurring filter. Hence, we rewrite Equation 3.46 as follows:

$$
\begin{aligned}
\widetilde{f}_{\text{BP}}(x, y) &= \mathcal{F}_2^{-2} \left(H' \cdot F \right)(x, y) \\
&= \int_0^\pi \mathrm{d}\phi \int_{-\infty}^{\infty} \mathrm{d}\nu \, H'(\nu, \phi) F(\nu, \phi) |\nu| e^{j2\pi\nu(x\cos\phi + y\sin\phi)} \\
&\stackrel{\text{CST}}{=} \int_0^\pi \mathrm{d}\phi \int_{-\infty}^{\infty} \mathrm{d}\nu \, H'(\nu, \phi) P(\nu, \phi) |\nu| e^{j2\pi\nu(x\cos\phi + y\sin\phi)}
\end{aligned}
$$

(3.47)

If we set

$$
H'(\nu, \phi) = \frac{1}{|\nu|}
$$

(3.48)

in Equation 3.47 then we get

$$
\begin{aligned}
\tilde{f}_{\mathrm{BP}}(x, y) &= \int_0^\pi \mathrm{d}\phi \int_{-\infty}^\infty \mathrm{d}\nu \, \frac{1}{|\nu|} P(\nu, \phi) \, |\nu| \, e^{j2\pi\nu(x\cos\phi + y\sin\phi)} \\
&= \int_0^\pi \mathrm{d}\phi \int_{-\infty}^\infty \mathrm{d}\nu \, P(\nu, \phi) \, e^{j2\pi\nu(x\cos\phi + y\sin\phi)} \\
&= \int_0^\pi \mathrm{d}\phi \, p(x\cos\phi + y\sin\phi, \phi)
\end{aligned}
$$

$$(3.49)$$

which demonstrates that the choice of the 2D blurring filter H' defined in Equation 3.48 is coherent with the definition of the laminogram in Equation 3.46. In other words, we have found out that by backprojecting the unfiltered sinogram we get an image given by the convolution of f with a 2D blurring filter with frequency response $1/\sqrt{u^2 + v^2} = 1/|\nu|$.

The identity in Equation 3.46 suggests another reconstruction method consisting of the 2D deblurring of the laminogram $\tilde{f}_{\mathrm{BP}}$. Deblurring in the frequency space is intuitively done by taking the reciprocal of the blurring filter H', with frequency response $\widehat{H}' = 1/H' = \sqrt{u^2 + v^2}$, and then writing

$$
\tilde{f}_{\mathrm{BPF}} = \mathcal{F}_2^{-1}(\widehat{H}' \cdot \tilde{F}_{\mathrm{BP}}) = \widehat{h}' ** \tilde{f}_{\mathrm{BP}} \, . \tag{3.50}
$$

Basically, an implementation of the BPF algorithm consists of switching the filtering and backprojection steps of the FBP algorithm, i.e., by first backprojecting the unfiltered sinogram in the image space and then deblurring the intermediate image by applying the 2D filter $\widehat{H}'$. Also in this case, 2D apodisation windows $A'(\nu, \phi)$ can be used in practice to control the image noise in the reconstructed image:

Algorithm 6 - BPF

---

1. Reconstruct the intermediate image $\widetilde{f}_{BP}$ (i.e., the laminogram of f), by backprojecting each row of the unfiltered sinogram p on the image plane along its own direction of projection ϕ, i.e., $\widetilde{f}_{BP} = \mathcal{R}_2^{\#} p$;

2. select the 2D apodisation window $A'(\nu, \phi)$ and compute the discrete version of the deblurring filter $A'\widehat{H}'$;

3. multiply the 2D deblurring filter $A'\widehat{H}'$ to the function $\widetilde{F}_{BP} = \mathcal{F}_2 \widetilde{f}_{BP}$ in the frequency space;

4. reconstruct the image by taking the inverse 2D DFT of the result of the previous point, i.e., $\widetilde{f}_{BPF} = \mathcal{F}_2^{-1}(A'\widehat{H}'\widetilde{F}_{BP})$.

3.6 SUGGESTED EXPERIMENTS

- Try to reproduce the results obtained in this chapter, starting from the Jupyter notebooks `DFRDemo`, `FBPParallelBeamDemo`, `FBPFanBeamDemo`, and `FDKDemo`.

- In the DFR demo, try to pad with zeros the phantom sinogram before taking its FFT. (Hint: use the *Numpy* function `pad` - see documentation at `https://docs.scipy.org/doc/numpy/reference/generated/numpy.pad.html`).

- In the parallel beam FBP demo, try to apodise the standard ramp filter by using the *Numpy* window functions (see documentation at `https://docs.scipy.org/doc/numpy/reference/routines.window.html`).

- In the parallel beam demo, try to change the interpolation method used in the backprojection step, and compare the RMS errors of the reconstructed image for each interpolation method.

- Try to modify the FDK demo, in order to reconstruct the full volume (or even just a few slices if the computer in use is not performant) on smaller arcs without angular subsampling, as is done for the fan beam exercises (see Figure 3.12).

Iterative reconstruction algorithms

As discussed in [38] all the iterative algorithms can be thought of as "closed loop" systems where the image estimate is updated at every iteration using an expression named the "update rule". The update rule of each iterative algorithm consists of mainly four elements (not necessarily in this order)

- Forward projection

- Comparison with experimental measure

- Back projection

- Application of normalization factors

- Update of the image estimate

We will highlight these elements in the code listings. In the following paragraphs, to describe more concisely the update rule, we will make use of the $q_k[\cdot]$, $\mathbf{q}[\cdot]$ $\mathbf{b}_k[\cdot]$, $\mathbf{b}[\cdot]$ operators that were described in Sections 1.5.3 and 1.5.4. All the iterative algorithms implemented in

DAPHNE use `IterativeReconstruction` as the base class. A derived class, i.e., a specific iterative algorithm, must implement the `EvaluateWeightingFactors` and `PerfomSingleIteration` methods. The `Evaluate-WeightingFactors` method triggers the calculation of the normalization factors that are needed during the update rule. Since the normalization factors do not depend on the measured data, they are calculated and stored prior to the start of the actual reconstruction. The `PerfomSingleIteration` method implements the update rule and contains a number of forward and back projection operations that can be invoked by calling `ForwardProjection` and back projection `Back-Projection` methods of the `IterativeReconstruction` base class. All the iterative algorithms need the following class attribute to be defined:

- `_image`: the current estimate of the image, i.e., the vector $\mathbf{f}$

- `_projection_data`: the experimental data to be reconstructed, i.e., the vector $\mathbf{p}$.

In this chapter we will focus on four of the most popular iterative algorithms: Algebraic Reconstruction Technique (ART), Simultaneous Iterative Reconstruction Technique (SIRT), Maximum Likelihood Expectation Maximization (MLEM), Order Subset Expectation Maximization (OSEM). A brief description of these algorithms is found in the next sections. For a more elaborate overview, the reader is referred to [20, 38, 26]. Finally, we would like to stress that algebraic algorithms work without any modification for both emission and transmission imaging. The original formulation of MLEM and OSEM instead fits emission data which theoretically follows the Poisson distribution. The adaptation of MLEM to transmission imaging is not included in DAPHNE and can be found in Elbakri and Fessler [5] or Nuyts *et al.* [22].

4.1 SYSTEM MATRIX

As discussed in Chapter 1 the SM is the most important ingredient of an iterative reconstruction method and, due to its footprint, is typically stored as a sparse matrix. There are several ways to compute the SM, ranging from analytical, to experimental and Monte Carlo methods. In real life, a trade-off between the image accuracy and the reconstruction time is always chosen. The description of these methods is outside the scope of this book.

We decided to implement in DAPHNE the "workhorse" of iterative reconstruction: the Siddon's ray tracing [29]. Siddon does not account for the physics effect other than solid angle. More complex physics models are often used in order to push the spatial resolution towards the intrinsic limit of the imaging device. In DAPHNE, the SM is calculated "on the fly": each row (i.e., each TOR) of the SM is calculated when needed. This allowed us to relax the memory requirements needed to run the examples and made the code easier to be understand. More details on the TOR computation are given in Chapter 5. A TOR can be computed using the method `ComputeTOR` of the `IterativeReconstruction` class. The `ComputeTOR` method (see listing 4.1) gives as output a Numpy rec-array, i.e. a Numpy ndarray whose columns can be accessed by name, of `TOR_dtype`. Each entry of the rec-array consists of four elements (`'vx'`, `'vy'`, `'vz'`, `'prob'`), where vx,vy,vz are the Cartesian indices of voxel, `prob` and its the probability of contributing to the projection.

```
1 from   Algorithms.IterativeReconstruction import
        IterativeReconstruction
2 ...
3 IterativeReconstruction it
4 # load the experimental setup of my reconstruction into
5 # the IterativeReconstruction class instance
```

```
6  it.SetExperimentalSetup(my_experimental_setup)
7  # compute the k-th row, i.e. the k-th TOR, of SM
8  MyTOR=it.ComputeTOR(k)
```

Listing 4.1: Code showing how to compute a TOR.

4.2   IMPLEMENTATION OF THE FORWARD AND BACK PROJECTION

As discussed in Chapter 1, both back and forward projection can be expressed as operations on $\mathbf{A}$ and $\mathbf{A}^t$ where each row, i.e. each TOR, of the SM is evaluated "on the fly". Using DAPHNE, a TOR can be evaluated invoking the `ComputeTOR` method of the `IterativeReconstruction` class (see listing 4.2).

```
1  from   Algorithms.IterativeReconstruction import
       IterativeReconstruction
2  ...
3  IterativeReconstruction it
4  # load the experimental setup of my reconstruction into
5  # the IterativeReconstruction class instance
6  it.SetExperimentalSetup(my_experimental_setup)
7  # compute the k-th row, i.e. the k-th TOR, of SM
8  MyTOR=it.ComputeTOR(k)
9  # forward projection of the TOR using the image img
10 # this operation implements eq 1.12
11 q=np.sum(img[MyTOR["vx"], MyTOR["vy"], MyTOR["vz"]] *
       MyTOR["prob"])
12 # allocate backprojection image
13 bp_image=np.zeros(img.shape);
14 # perform backprojection of q
15 # this operation implements eq 1.14
16 bp_image[MyTOR["vx"], MyTOR["vy"], MyTOR["vz"]] += q *
       MyTOR["prob"]
```

Listing 4.2: Code showing the implementation of forward and back projections.

4.3   HADAMARD PRODUCT AND DIVISION

In this chapter we will make use of two vector operations that are not met frequently: the Hadamard product ∘, and the Hadamard division ⊘. Even if these two names could frighten the reader, they express only the element-wise product and division between two vectors of the same length. For example, given $\mathbf{a} = (a_1, .., a_N)$ and $\mathbf{b} = (b_1, .., b_N)$ we denote

$$\mathbf{a} \circ \mathbf{b} = (a_1 \cdot b_1, .., a_N \cdot b_N) \qquad (4.1)$$

and

$$\mathbf{a} \oslash \mathbf{b} = (a_1/b_1, .., a_N/b_N). \qquad (4.2)$$

4.4 ALGEBRAIC RECONSTRUCTION TECHNIQUE (ART)

ART was originally proposed in [9] and it is an implementation of the Kaczmarz's method for solving linear systems. This reconstruction technique does not make any hypothesis on the underlying nature of the reconstructed data, i.e., it is purely algebraic. The ART considers Equation 1.11 as a linear system describing the intersection of M hyperplanes, where the equation of the k-th hyperplane is given by

$$A_{k,1}f_1 + A_{k,2}f_2 + .. + A_{k,N}f_N = p_k. \qquad (4.3)$$

If existent, the N-dimensional point $\mathbf{f}^* = (f_1^*, .., f_N^*)$ where all the M hyperplanes intersect is the solution of the linear system, i.e. the searched image. An approximate solution of Equation 1.11 can be found by projecting geometrically the current image estimate onto every hyperplane (see Figure 4.1). Note that, in case of a M-dimensional space and M orthogonal hyperplanes, the exact solution is reached in at most M iterations. On the other hand, in order to boost the convergence speed, we should avoid processing almost parallel hyperplanes consecutively. It is important to realize that when using ART, the estimate is updated after each projection is processed. For this reason we will denote each image estimate $\mathbf{f}_k^n$ using two indices: the iteration index (n) and the

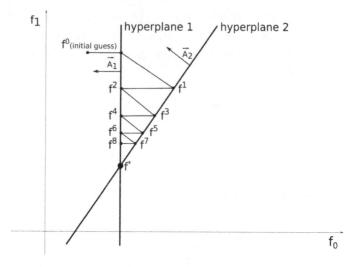

Figure 4.1: Picture showing the idea behind the update rule of the ART algorithm, when N=M=2.

index of the projection (k). We will increment the iteration index only when the algorithm has gone through all the M projections of the system (see listing 4.3 and algorithm 7). The typical choice to start the ART iterative procedure is to set all the voxels of the image matrix to $\mathbf{f}_0^0 = \mathbf{0}$. Each time we process a projection, the estimate is updated according to

$$\mathbf{f}_{k+1}^n = \mathbf{f}_k^n - \mathbf{A}_k^t((\mathbf{A}_k\mathbf{f}_k^n - p_k)(L_k)) \qquad (4.4)$$

where A_k are the row of the SM and

$$L_k = \frac{1}{\sum\limits_{i=0}^{N} A_{k,i}^2}. \qquad (4.5)$$

ART is known to converge in a low number of iterations. However, images reconstructed with this algorithm suffer from artifacts due to inconsistencies in real projection data.

Algorithm 7 - ART

for k in $1, .., M$ do

1. Forward projection: compute $q_k[\mathbf{f}_k^n]$

2. Comparison with experimental data: compute $\Delta_k = q_k - p_k$

3. Application of normalization factors: compute $Z_k = \Delta_k \cdot L_k$

4. Back projection: compute $\mathbf{b}_k[Z_k]$ and store it in $\mathbf{I}$

5. Update of the image estimate: subtract $\mathbf{I}$ from $\mathbf{f}_k^n$ and store the result in $\rightarrow \mathbf{f}_{k+1}^n$

end for

```
1   def PerfomSingleIteration(self):
2       for l in range(self.GetNumberOfProjections()):
3           # forward projection
4           proj = self.ForwardProjectSingleTOR(self.
    _image, l)
5           # comparison with expertimental measures (
    subtraction)
6           proj -= self._projection_data[l]
7           # application normalization factors
8           proj *= self._L[l]
9           # backprojection and update current estimate
10          self.BackProjectionSingleTOR(self._image, -
    proj, l)
```

Listing 4.3: Code showing the update rule of the ART algorithm (listing of code from Algorithms/ART.py).

4.5 SIMULTANEOUS ITERATIVE RECONSTRUCTION TECHNIQUE (SIRT)

Simultaneous Iterative Reconstruction Technique was originally described in [8] and can be considered a variant of ART designed to produce smoother images. This feature comes at the expense of a slower convergence. The reason for this behaviour is that, unlike ART, the image is updated only once every projection is processed. As for the ART, the most common choice to start the iterative procedure is to set $\mathbf{f}^0 = \mathbf{0}$.

The update rule for SIRT is:

$$\mathbf{f}^{n+1} = \mathbf{f}^n + \mathbf{C} \circ \mathbf{A}^t \circ \mathbf{R} \circ (\mathbf{p} - \mathbf{A} \cdot \mathbf{f}^n) \qquad (4.6)$$

where $\mathbf{C} = 1 \oslash (\sum_{i=0}^{N} A_{i,j}) \in \mathbb{R}^M$ and $\mathbf{R} = 1 \oslash (\sum_{j=0}^{M} A_{i,j}) \in \mathbb{R}^N$ are vectors containing respectively the reciprocal of the sum over columns and rows of the system model. The code of the update rule and the pseudocode are shown respectively in listing 4.4 and algorithm 8.

```
1  def PerfomSingleIteration(self):
2          # forward projection
3          proj = self.ForwardProjection(self._image)
4          # comparison with experimental measures (
   subtraction)
5          proj -= self._projection_data
6          # application normalization factors
7          proj *= self._C
8          # backprojection and update current estimate
9          self._image = self.BackProjection(-proj)* self.
   _R
```

Listing 4.4: Code showing the update rule for the SIRT algorithm (listing of code from Algorithms/SIRT.py).

Algorithm 8 - SIRT

---

1. Forward projection: compute $\mathbf{q}[\mathbf{f}^n]$

2. Comparison with experimental data: compute $\boldsymbol{\Delta} = \mathbf{p} - \mathbf{q}$

3. Application of normalization factors: compute $\mathbf{Z} = \boldsymbol{\Delta} \circ \mathbf{R}$

4. Back projection: compute $\mathbf{b}[\mathbf{Z}]$ and store it in $\mathbf{I}$

5. Application of normalization factors: compute $\mathbf{D} = \mathbf{C} \circ \mathbf{I}$

6. Update of the image estimate: add $\mathbf{D}$ to $\mathbf{f}^n$ and store the result in $\rightarrow \mathbf{f}^{n+1}$

4.6 MAXIMUM-LIKELIHOOD EXPECTATION MAXIMIZATION (MLEM)

MLEM was originally described in [34] and it is one of the most popular iterative reconstruction methods nowadays. The MLEM works under the hypothesis that the acquired data follows the Poisson distribution. This is the typical situation in emission tomography. Given that the j-th detector element counts follow a Poisson distribution with unknown average $\bar{p}_j$, the likelihood function of a set of the detector is

$$\mathcal{L}(\bar{p}_1, .., \bar{p}_M | f_1, .., f_N) = \prod_{j=1}^{M} \frac{\bar{p}_j^{n_j}}{n_j!} e^{-\bar{p}_j} \qquad (4.7)$$

where $\bar{p}_j = \sum_{i=0}^{N} A_{j,i} \cdot f_i$. The likelihood can be maximized (ML) finding the set of f_i such that

$$\frac{\partial \ln(\mathcal{L}(f_1, ..f_N))}{\partial f_i} = 0. \qquad (4.8)$$

After working out the mathematics and imposing the "estimation maximization step" [38], the following update rule is found:

$$\mathbf{f}^{n+1} = \mathbf{f}^n \oslash \mathbf{S} \circ \mathbf{A}^t \left(\mathbf{p} \oslash (\mathbf{A} \cdot \mathbf{f}^n) \right) \qquad (4.9)$$

where

$$\mathbf{S} = \mathbf{A}^t \cdot \mathbf{1} \qquad (4.10)$$

is often referred to as the sensitivity matrix. The update rule is also shown in listing 4.5 and algorithm 4.5. The typical choice for starting the MLEM iterative procedure is to set $\mathbf{f}^0 = \mathbf{1}$.

```
1   def PerfomSingleIteration(self):
2       # forward projection
3       proj = self.ForwardProjection(self._image)
4       # this avoid 0 division
5       nnull = proj != 0
6       # comparison with experimental measures (ratio)
7       proj[nnull] = self._projection_data[nnull] /
    proj[nnull]
8       # backprojection
9       tmp = self.BackProjection(proj)
10      # apply sensitivity correction and update
    current estimate
11      self._image = self._image * self._S * tmp
```

Listing 4.5: Code showing the update rule for the MLEM algorithm (listing of code from Algorithms/MLEM.py).

MLEM is known to produce high quality images; however it suffers from a slow convergence.

4.7 ORDERED-SUBSET EXPECTATION MAXIMIZATION (OSEM)

Ordered Subset Expectation Maximization was originally described in [12]. It can be considered a generalization of MLEM aimed to speed up the convergence. It consists of dividing the experimental measurements in a number of non overlapping subsets, $\mathbf{p}_1$, .., $\mathbf{p}_l$ and then running the MLEM algorithm on each subset sequentially, so that the reconstruction of the i+1-th subset starts from the estimate given by the i-th. The Update rule of OSEM is given by:

$$\mathbf{f}_l^n = \mathbf{f}_{l-1}^n \oslash \mathbf{S}_l \circ \mathbf{A}^T \left(\mathbf{p_l} \oslash \left(\mathbf{A} \cdot \mathbf{f}_{l-1}^n \right) \right) \tag{4.11}$$

and where S_l is the sensitivity of the l-th subset,i.e.,

$$\mathbf{S}_l = \mathbf{A}^t \cdot \mathbf{I}_l \tag{4.12}$$

with

$$I_l = \left\{ \begin{array}{l} 1, \text{ if projection} \in l - \text{th subset} \\ 0, \text{ otherwise} \end{array} \right\} \tag{4.13}$$

One iteration is defined when the algorithm has gone through all the subsets which are visited always in the

---

Algorithm 9 - MLEM

---

1. Forward projection: compute $\mathbf{q}[\mathbf{f^n}]$;

2. Comparison with experimental data: compute $\mathbf{\Delta} = \mathbf{p} \oslash \mathbf{q}$;

3. Back projection: compute $\mathbf{b}[\mathbf{\Delta}]$ and store the result in $\mathbf{I}$;

4. Application of normalization factors: compute $\mathbf{I} \circ \mathbf{f}^n \oslash \mathbf{S}$ and store it in $\mathbf{I}$;

5. Update of the image estimate: $\mathbf{I} \to \mathbf{f}^{n+1}$

---

same order (see listing 4.6 and algorithm 10). Using OSEM the convergence of the MLEM is sped-up by a factor equal to the number of subsets used. For example, if the reconstruction of some data using MLEM takes 100 iterations to reach a certain image quality, it will get roughly 20 iterations of OSEM with 5 subsets to obtain the same results. However, using OSEM, the convergence to the maximum-likelihood estimator is not guaranteed and what is observed is that OSEM cycles between slightly different image estimates [33]. The choice of the subset population is therefore of paramount importance. In DAPHNE, projection data are divided into subsets randomly, to favor an even distribution of sampling.

```python
def PerfomSingleSubsetIteration(self, subset):
        # Get all the projections belonging to the
        current subset
        LORCurrentSubset = self._subsets[subset]
        # perform the forward projection using only
        projection
        # belonging to the current subset
        proj = self.ForwardProjection(self._image,
        LORCurrentSubset)
        # avoid 0 division
        inv_proj = np.zeros_like(proj, projection_dtype)
        nnull = proj != 0
        # comparison with experimental measures of the
        current subset (ratio)
        inv_proj[nnull] = self._data_subsets[subset][
        nnull] / proj[nnull]
        # perform the forward projection using only
        projection
        # belonging to the current subset
        tmp = self.BackProjection(inv_proj,
        LORCurrentSubset)
        # apply sensitivity correction for current
        subset and update current estimate
        self._image = self._image * self._S[subset] *
        tmp

def PerfomSingleIteration(self):
    for subset in range(self._subsetnumber):
        self.PerfomSingleSubsetIteration(subset)
```

Listing 4.6: Code showing the update rule for the OSEM algorithm (listing of code from Algorithms/OSEM.py).

Algorithm 10 - OSEM

---

for i in $1, .., l$ do

1. Select the data of the i-th subset (p_i);

2. Perform an iteration of MLEM using p_i, S_i and starting from the image estimate f_{i-1}^n;

3. Store the result in f_i^n;

end for

4.8 A STEP-BY-STEP EXAMPLE USING ARTIFICIAL NOISELESS PROJECTION DATA

In this section we will show how to perform iterative reconstruction using DAPHNE. A code implementing all the described operations can be found in the folder `NotebookIterativeReconstructionDemoPET.ipynb`. Let's assume that we are dealing with a PET detector made of one slice. Working with such a simple experimental setup will allow us to perform a series of experiments in a reasonable time. However, we would like to stress that the code provided in DAPHNE is fully 3D and we encourage the reader to experiment with it. Prior to starting the reconstruction, we need to import a bunch of packages, as shown in listing 4.7.

```
1  import numpy as np # numpy
2  import matplotlib.pyplot as plt # matplotlib
3  from Misc.Utils import ReadImage
4  from Geometry.ExperimentalSetupPET import
       ExperimentalSetupPET
5  from Algorithms.ProjectionDataGenerator import
       ProjectionDataGenerator
6  from Algorithms.ART import ART
7  from Algorithms.SIRT import SIRT
8  from Algorithms.MLEM import MLEM
9  from Algorithms.OSEM import OSEM
10 from Misc.DataTypes import voxel_dtype
```

Listing 4.7: Step-by-step example: import phase.

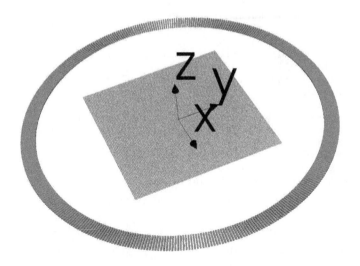

Figure 4.2: Picture showing a rendering of the generated experimental setup.

Then, as discussed in Chapter 2, we use the class ExperimentalSetupPET to define a cylindrical PET setup. A rendering like the one shown in Figure 4.2 can be obtained invoking, after the method Update is called, my_experimental_setup.Draw() method. The parameter my_experimental_setup.pixel_size defines the size of the pixel in mm, and is used for visualization purposes only and this parameter does not affect reconstruction results.

```
1 # create a PET experimental setup
2 my_experimental_setup = ExperimentalSetupPET()
3 # radius of the cylindrical PET
4 my_experimental_setup.radius=100
5 # size of the pixels (visualization purpose only)
6 my_experimental_setup.pixel_size=np.array([1,1,10])
7 # number of pixels in the cylindrical geom
8 my_experimental_setup.pixels_per_slice_nb=400
9 # number of slices of the PET detector
10 my_experimental_setup.detector_slice_nb=1
11 # slice pitch
12 my_experimental_setup.slice_pitch_mm=1
13 # image matrix size in mm
```

```
14  my_experimental_setup.image_matrix_size_mm = np.array
        ([100,100,100])
15  # voxel size in mm
16  my_experimental_setup.voxel_size_mm = np.array
        ([1,1,100])
17  # h size of for the coincidences
18  my_experimental_setup.h_fan_size = 80
19  # name of the experimental setup
20  my_experimental_setup.detector_name = "my first PET"
21  # perform all the calculations, e.g. place the pixels
        into the physical space
22  my_experimental_setup.Update()
23  # print some info about the generated experimental setup
24  print(my_experimental_setup.GetInfo())
```

Listing 4.8: Step-by-step example: geometry definition.

After defining the setup, we can generate artificial projection data using the `ProjectionDataGenerator` class (see listing 4.9). For this example, we will use the 2D Shepp Logan Phantom contained in the `Data` folder. Note that the phantom image must have the same size of the image matrix, i.e., `my_experimental_setup._voxel_nb`. In this case the `SheppLoganPhantom_100x100x1.png` matches the 100x100x1 size of the image matrix size defined in the example. The `Projection-DataGenerator` class calculates the forward projection of `input_img` using the `_projections_extrema` defined in `my_experimental_setup` and returns them in np.array.

```
1  # load the shepp long image to generate projections
2  input_img = ReadImage("SheppLoganPhantom_100x100x1.png")
3  plt.imshow(np.squeeze(input_img))
4  g=ProjectionDataGenerator(my_experimental_setup)
5  # add noise to proj: 0 no noise 1 poisson noise 2
        gaussian noise
6  noise=0
7  # this calculates the forward projection of input_img
        using the
8  # projection extrema contained in my_experimental_setup
9  projections=g.GenerateObjectProjectionData(input_img,
        noise,0)
```

Listing 4.9: Step-by-step example: generation of projections.

In this example we will work with noiseless projection. In order to simulate statistical fluctuations, in the next section we will work with projections affected by Poisson noise. Once the projections are defined we can start with the actual reconstruction. In this example we will use ART even though the reader can easily switch to SIRT, MLEM or OSEM by changing the variable algorithm. The piece of code shown in listing 4.10 will perform 10 ART iterations starting from an image f^0 filled with zeros. Note that if you plan to use MLEM/OSEM you should set initial_value to some non null number, as these algorithms apply multiplicative corrections to the original image (see MLEM and OSEM update rules). After the reconstruction is run, you should get results similar to those shown in Figure 4.3. This example shows that, in the case of noiseless projections, ART converges in a low number of iterations to an image very close to the real solution of the problem. However, when projections data are affected by statistical noise the situation can be different. Moreover, the user should consider that even in the case of noiseless data, a good reconstruction can be achieved only if a sufficient sampling of the image matrix is available, i.e. there must be a sufficient number of non collinear projections passing through each voxel of the image matrix. To familiarize yourself with this concept can try to reduce the pixels_per_slice_nb and h_fan_size variables.

```
1  # algorithm must be one of "MLEM", "ART", "SIRT", "OSEM"
2  algorithm="ART"
3  # number of iterations
4  niter=10
5  # number of subsets (OSEM only)
6  nsubsets=5
7  # uniform value to be set into the image_matrix, i.e. f
        ^0
8  initial_value=0
9  # launch the constructor of the algorithm
10 # defined in the var algorithm
11 it = eval( algorithm+ "()")
12 it.SetExperimentalSetup(my_experimental_setup)
13 # number of iterations to be performed
14 it.SetNumberOfIterations(niter)
15 # number of subsets used only when OSEM is called
```

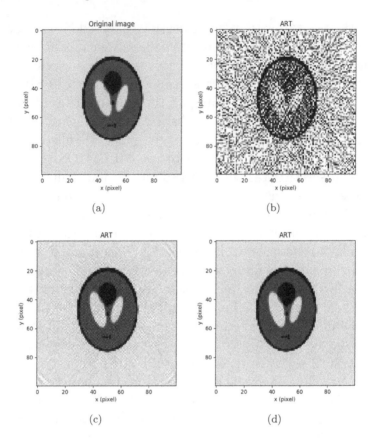

Figure 4.3: Original image a) and images reconstructed using ART algorithm in the case of noiseless projections after 1 iteration b), 5 iterations c), 10 iterations d).

```
16  it.SetNumberOfSubsets(nsubsets)
17  # the artificial projections evaluated in the previous
        step
18  it.SetProjectionData(projections)
19  # start with a initial_guess filled image
20  initial_guess=np.full(it.GetNumberOfVoxels(),
        initial_value, dtype=voxel_dtype)
21  it.SetImageGuess(initial_guess)
22  # perform the reconstruction and return the last
        iteration
23  output_img = it.Reconstruct()
```

Listing 4.10: Step-by-step example: reconstruction.

4.9 A STEP-BY-STEP EXAMPLE USING ARTIFICIAL POISSON NOISE AFFECTED DATA

As discussed in the previous section, noise affecting projections can degrade substantially the quality of the reconstructed images. In this section we will describe an experiment to assess the algorithm performance when dealing with projections affected by Poisson noise. The only thing that we need to do in order to do that is to change `noise` variable to 1 in the listing 4.9. This would cause the `ProjectionDataGenerator` to extract the values of the projections from a Poisson random distribution, with each distribution having an average equal to the noiseless projection. Note that the Poisson distribution is discrete, i.e. the extracted values can be only integers. This means that if the average of a distribution «1 almost all the values extracted will be 0. To avoid this issue, check the values of the projections before turning on the noise option. If we repeat the same experiment of the previous section with Poisson affected data we obtain the images shown in Figure 4.4

4.9.1 Study of convergence properties of the algorithms

As discussed before iterative algorithms display different behavior both in terms of convergence speed and image quality. To convince ourselves we will repeat systematically the experiments shown in the previous sections for all the algorithms proposed in this book: ART, SIRT, MLEM and OSEM with 5 subsets. The recipe that we are going to apply is the following:

- for each algorithm we run 100 iterations

- for each iteration and algorithm we evaluate the root mean square error (RMS) normalized to the maximum value of the image (this is to minimize the dependence on the actual image values)

- we plot RMS vs. iteration for each algorithm

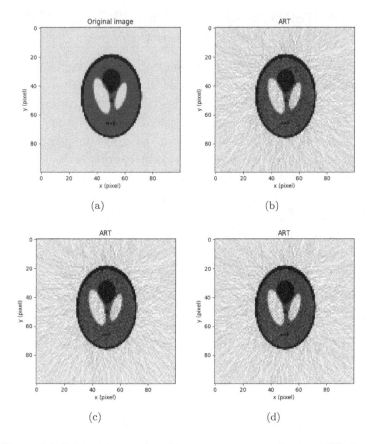

Figure 4.4: Original image a) and images reconstructed using ART algorithm in the case of Poisson noise affected projections after 5 iterations b), 10 iterations c), 15 iterations d).

Moreover, we will perform the experiment both with noiseless and Poisson affected projections. The plots of RMS vs. the iterations, in the case of noiseless and Poisson affected projections are shown respectively in Figure 4.5 and Figure 4.6. As is clear from the plot, ART performs quite well in the case of noiseless data: it converges in a couple of iterations to a very close approximation of the original image. SIRT reconstructs a smoother image; however this is paid for in terms of slow convergence.

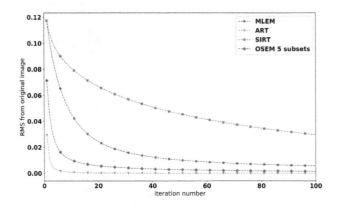

Figure 4.5: Plot of the RMS vs. iteration for MLEM, ART, SIRT, OSEM 5 subsets in the case of noiseless data.

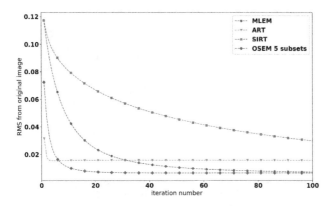

Figure 4.6: Plot of the RMS vs. iteration for MLEM, ART, SIRT, OSEM 5 subsets in the case of Poisson noise affected data.

MLEM and OSEM converge to the same RMS of ART but in a much higher number of iterations. For what concerns the Poisson affected projections, the situation is different: SIRT is still the slower algorithm to converge, ART converges very quickly; however, MLEM and OSEM reach a lower RMS. For explaining this behaviour

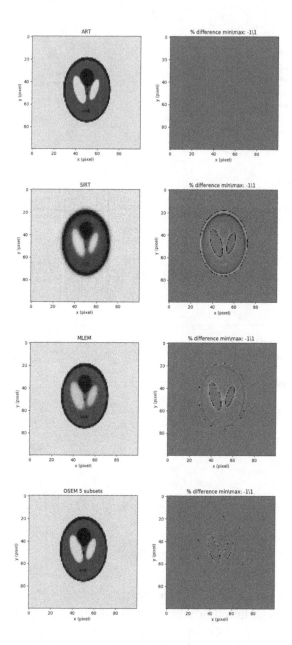

Figure 4.7: Result of the experiment with noiseless projections. Left column: images reconstructed after 50 iterations with ART, SIRT, OSEM and MLEM. Right column: relative difference with respect to the original image.

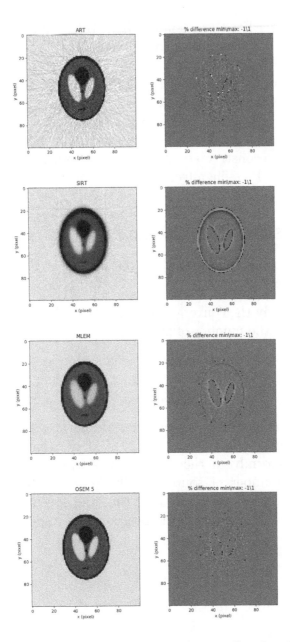

Figure 4.8: Result of the experiment with Poisson affected projections. Left column: images reconstructed after 50 iterations with ART, SIRT, OSEM and MLEM. Right column: relative difference with respect to the original image.

the reader has to remember that MLEM and OSEM are formulated for Poisson distributed data whereas ART and SIRT do not model the statistics distribution. This is of course a limited experiment and the RMS is not the only metric to assess image similarity; nonetheless it gives us an idea of the heuristic behavior of the discussed algorithms. Finally, we can have a look at the images reconstructed at the 50-th iteration for noiseless and Poisson affected projections (see Figures 4.7 and 4.8).

4.10 SUGGESTED EXPERIMENTS

- Try to reproduce the experiments of the previous sections.

- Study the convergence behavior of the OSEM varying the number of subsets.

- Non linearity: design a small experiment to show that the MLEM is non linear according to the definition of Chapter 1.3.1.

- Non uniform convergence: design a small experiment to show that the MLEM does not converge uniformly in the image matrix. Suggestion: place a number of sources in different positions of the image matrix and show that the speed of convergence depends on the position of the source.

## CHAPTER 5

Overview of methods for generation of projection data

5.1 ANALYTICAL PROJECTION OF IDEAL ELLIPSOIDAL PHANTOMS

In most cases, complex objects of interest for PET/CT imaging can be modeled as just the superposition of ellipsoids with constant density or activity concentration. The generation of useful projection data for these objects can be done by knowing:

- the length of each of the three axes of the ellipsoid (3 parameters),

- the center of the ellipsoid in a given frame of reference (3 parameters),

- the angles defining the orientation of the ellipsoid in the given frame of reference (3 parameters),

- the parameters defining a line of projection in the same frame of reference (i.e., the two endpoints or the source point and direction vector) (6 parameters).

Given the information above (15 parameters), the projection data can be computed by starting from the intersection length of the selected line of projection with the ellipsoid, and then by multiplying this length by its constant density/activity concentration. The following derivation is adapted from Ref. [7].

In order to calculate the intersection length, let us consider an $Oxyz$ orthogonal reference frame, and let us denote by $\mathbf{p} = (x, y, z)$ a point in the 3D space. Let us now consider an ellipsoid with the three semi-axes of length a, b and c, respectively: by defining the diagonal matrix

$$M = \begin{pmatrix} \frac{1}{a} & 0 & 0 \\ 0 & \frac{1}{b} & 0 \\ 0 & 0 & \frac{1}{c} \end{pmatrix} \tag{5.1}$$

the ellipsoid can be defined conveniently by the implicit equation

$$\mathbf{p}M \cdot \mathbf{p}M = \frac{x^2}{a^2} + \frac{y^2}{b^2} + \frac{z^2}{c^2} = 1. \tag{5.2}$$

If the center of the ellipsoid is located elsewhere than on the origin O, let's say at the point $\mathbf{c} = (x_c, y_c, z_c)$, the above equation can be generalized as follows:

$$(\mathbf{p}M - \mathbf{c}M) \cdot (\mathbf{p}M - \mathbf{c}M)$$
$$= \frac{(x - x_c)^2}{a^2} + \frac{(y - y_c)^2}{b^2} + \frac{(z - z_c)^2}{c^2} = 1. \tag{5.3}$$

Let us now define a line of projection, through its source point $\mathbf{p_0} = (x_0, y_0, z_0)$ and its direction vector $\mathbf{v} = (v_x, v_y, v_z)$, with $|v|^2 = 1$. The line can be expressed by a parametric equation:

$$\mathbf{p} = \mathbf{p_0} + t\mathbf{v}. \tag{5.4}$$

The intersection of the line and the ellipsoid is found by solving simultaneously the set of the two Equations 5.3

and 5.4 with respect to t. By defining the two auxiliary vectors $\mathbf{p_1} = \mathbf{p_0}M - \mathbf{c}M$ and $\mathbf{v_1} = \mathbf{v}M$, the two solutions of the system of equations are:

$$t_{1,2} = \frac{-\mathbf{p_1} \cdot \mathbf{v_1} \pm \sqrt{(\mathbf{p_1} \cdot \mathbf{v_1})^2 - |v_1|^2 (|p_1|^2 - 1)}}{|v_1|^2} \quad (5.5)$$

and hence, by replacing in Equation 5.4, we've found the two intersection points

$$\mathbf{p}_{1,2} = \mathbf{p_0} + t_{1,2}\mathbf{v}. \quad (5.6)$$

The above result can be generalized to the case of an arbitrarily oriented ellipsoid, i.e., when the three semi-axes are not parallel to the main axes of the $Oxyz$ frame. To do this, let us define a new, rotated reference frame $O'x'y'z'$ in which the ellipsoid is centered at the origin with its axes parallel to x', y' and z'.

The transformation between the two reference frames is done by

$$\mathbf{p'} = \begin{pmatrix} x' \\ y' \\ z' \end{pmatrix} = R(\alpha_x, \alpha_y, \alpha_z) \begin{pmatrix} x - x_c \\ y - y_c \\ z - z_c \end{pmatrix} \quad (5.7)$$

with $R(\alpha_x, \alpha_y, \alpha_z)$ being the composition of the three rotation matrices around the main axes x, y and z (in general, in reversed order):

$$R(\alpha_x, \alpha_y, \alpha_z) = R_z(\alpha_z)R_y(\alpha_y)R_x(\alpha_x), \quad (5.8)$$

where

$$R_x(\alpha_x) = \begin{pmatrix} 1 & 0 & 0 \\ 0 & \cos\alpha_x & \sin\alpha_x \\ 0 & -\sin\alpha_x & \cos\alpha_x \end{pmatrix}, \quad (5.9)$$

$$R_y(\alpha_y) = \begin{pmatrix} \cos\alpha_y & 0 & -\sin\alpha_y \\ 0 & 1 & 0 \\ \sin\alpha_y & 0 & \cos\alpha_y \end{pmatrix}, \quad (5.10)$$

$$R_z(\alpha_z) = \begin{pmatrix} \cos\alpha_z & \sin\alpha_z & 0 \\ -\sin\alpha_z & \cos\alpha_z & 0 \\ 0 & 0 & 1 \end{pmatrix}. \tag{5.11}$$

Using the transformation defined in Equation 5.7, the equation of the ellipsoid becomes

$$\mathbf{p}'M \cdot \mathbf{p}'M \cdot = 1, \tag{5.12}$$

whereas the line of projection is written as

$$\mathbf{p}' = \mathbf{p}'_0 + t\mathbf{v}', \tag{5.13}$$

with $\mathbf{v}' = R(\mathbf{v} - \mathbf{c})$ and

$$\mathbf{p}'_0 = \begin{pmatrix} x'_0 \\ y'_0 \\ z'_0 \end{pmatrix} = R \begin{pmatrix} x_0 - x_c \\ y_0 - y_c \\ z_0 - z_c \end{pmatrix}. \tag{5.14}$$

As done previously, the intersection points of the line and ellipsoid in the rotated frame are just computed by solving the system of the two Equations 5.12 and 5.13 for the unknown t. If we define again two auxiliary vectors $\mathbf{p}'_1 = \mathbf{p}'_0 M$ and $\mathbf{v}'_1 = \mathbf{v}'M$, the two solutions for t have identical form as those obtained in Equation 5.5 for the non-rotated frame. It is obvious that, if the quantity inside the square root in Equation 5.5 is negative, there will be no intersection between the line and the ellipsoid; if instead that quantity is null, the line is tangent to the ellipsoid surface and we will have just one point of intersection.

5.2 NUMERICAL PROJECTION OF VOXELIZED PHANTOMS

5.2.1 Siddon's Algorithm

One of the most common problems faced when working in the image reconstruction field is to find the intersection between the image matrix and a ray connecting two

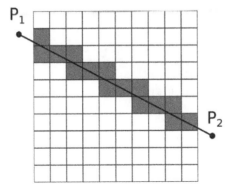

Figure 5.1: Ray tracing of the line connecting the physical points P_1 and P_2. The voxels colored in gray are those intersected by the ray.

physical points. This operation is called ray tracing and there are plenty of algorithms to perform that. One of the most commonly used in image reconstruction is the Siddon's [29]. Siddon's algorithm allows us to evaluate, at the same time, all the voxels intersected by a specific ray and their respective intersection length (see Figure 5.1). An implementation is included in DAPHNE in the file SiddonProjector.py and follows closely the algorithm described in the original paper. In DAPHNE Siddon's algorithm is used both to generate a set of simulated projections and to evaluate on the fly system matrix for the iterative reconstruction. An example of how the Siddon can be invoked by the user is shown in listing 5.1.

```
1  from Algorithms.SidddonProjector import SiddonProjector
2  from Misc.DataTypes import point_dtype
3  # set the size in mm of the image matrix
4  image_matrix_size_mm=np.array([100 ,100 ,100])
5  # set voxel size
6  voxel_size_mm=np.array([1, 1,100])
7  # create a SiddonProjector object
8  s=SiddonProjector(image_matrix_image_mm ,voxel_size_mm)
9  # define the ray extrema
10 P1=np.array((100,100,0),dtype=point_dtype)
11 P2=np.array((-100,-100,0),dtype=point_dtype)
```

```
12  # evaluate the intersection between the line connecting
         P1,P2
13  # and the image_matrix and display it
14  # if no image is displayed the ray does not intersect
15  # the FOV
16  TOR=s.DisplayLine(P1,P2)
```

Listing 5.1: Code showing how to use the SiddonProjector class file.

Bibliography

[1] H. O. Anger. Use of a Gamma-Ray Pinhole Camera for in vivo studies. *Nature*, 170(4318):200–201, August 1952. Number: 4318 Publisher: Nature Publishing Group.

[2] Mario Bertero and Patrizia Boccacci. *Introduction to Inverse Problems in Imaging*. CRC press, 1998.

[3] Tim Day. Image processing - What's wrong with this code for tomographic reconstruction by the Fourier method? Library Catalog: dsp.stackexchange.com.

[4] A. Del Guerra, N. Belcari, and M. Bisogni. Positron Emission Tomography: Its 65 years. *Nuovo Cimento Rivista Serie*, 39:155–223, April 2016.

[5] I.A. Elbakri and J.A. Fessler. Statistical image reconstruction for polyenergetic X-ray computed tomography. *IEEE Transactions on Medical Imaging*, 21(2):89–99, February 2002. Conference Name: IEEE Transactions on Medical Imaging.

[6] L.A. Feldkamp, L.C. Davis, and J.W. Kress. Practical cone-beam algorithm. *JOSA A*, 1(6):612–619, 1984.

[7] Bob Geitz. Vector Geometry for Computer Graphics. Available at https://cs.oberlin.edu/~bob/cs357.08/VectorGeometry/VectorGeometry.pdf.

[8] Peter Gilbert. Iterative methods for the three-dimensional reconstruction of an object from projections. *Journal of Theoretical Biology*, 36(1):105 – 117, 1972.

[9] Richard Gordon, Robert Bender, and Gabor T. Herman. Algebraic reconstruction techniques (art) for three-dimensional electron microscopy and x-ray photography. *Journal of Theoretical Biology*, 29(3):471 – 481, 1970.

[10] Pierre Grangeat. Mathematical framework of cone beam 3d reconstruction via the first derivative of the radon transform. In *Mathematical Methods in Tomography*, pages 66–97. Springer, 1991.

[11] Gabor T. Herman and Abraham Naparstek. Fast image reconstruction based on a radon inversion formula appropriate for rapidly collected data. *SIAM Journal on Applied Mathematics*, 33(3):511–533, 1977.

[12] H. M. Hudson and R. S. Larkin. Accelerated image reconstruction using ordered subsets of projection data. *IEEE Transactions on Medical Imaging*, 13(4):601–609, Dec 1994.

[13] A. Iriarte, R. Marabini, S. Matej, C.O.S. Sorzano, and R.M. Lewitt. System models for pet statistical iterative reconstruction: A review. *Computerized Medical Imaging and Graphics*, 48:30 – 48, 2016.

[14] Avinash C. Kak and Malcolm Slaney. *Principles of Computerized Tomographic Imaging*. IEEE Press, 1988.

[15] Willi A Kalender, Wolfgang Seissler, Ernst Klotz, and Peter Vock. Spiral volumetric ct with single-breath-hold technique, continuous transport, and

continuous scanner rotation. *Radiology*, 176(1):181–183, 1990.

[16] Alexander Katsevich, Samit Basu, and Jiang Hsieh. Exact filtered backprojection reconstruction for dynamic pitch helical cone beam computed tomography. *Physics in Medicine and Biology*, 49(14):3089, 2004.

[17] Thomas Kluyver, Benjamin Ragan-Kelley, Fernando Pérez, Brian Granger, Matthias Bussonnier, Jonathan Frederic, Kyle Kelley, Jessica Hamrick, Jason Grout, Sylvain Corlay, Paul Ivanov, Damián Avila, Safia Abdalla, Carol Willing, and Jupyter development team. Jupyter notebooks - a publishing format for reproducible computational workflows. In Fernando Loizides and Birgit Scmidt, editors, *Positioning and Power in Academic Publishing: Players, Agents and Agendas*, pages 87–90. IOS Press, 2016.

[18] Hiroyuki Kudo and Tsuneo Saito. Derivation and implementation of a cone-beam reconstruction algorithm for nonplanar orbits. *IEEE Transactions on Medical Imaging*, 13(1):196–211, 1994.

[19] Clemens Maaß, Frank Dennerlein, Frédéric Noo, and Marc Kachelrieß. Comparing short scan ct reconstruction algorithms regarding cone-beam artifact performance. In *IEEE Nuclear Science Symposuim & Medical Imaging Conference*, pages 2188–2193. IEEE, 2010.

[20] F. Natterer and F. Wübbeling. *Mathematical Methods in Image Reconstruction*. Society for Industrial and Applied Mathematics, 2001.

[21] Frank Natterer. *The Mathematics of Computerized Tomography*, Volume 32. Siam, 1986.

[22] Johan Nuyts, Bruno De Man, Jeffrey A. Fessler, Wojciech Zbijewski, and Freek J. Beekman. Modelling the physics in the iterative reconstruction for transmission computed tomography. *Phys Med Biol*, 58(12):R63–96, June 2013.

[23] D. Panetta, N. Belcari, A. Del Guerra, and S. Moehrs. An optimization-based method for geometrical calibration in cone-beam ct without dedicated phantoms. *Physics in Medicine and Biology*, 53(14):3841, 2008.

[24] Dennis L. Parker. Optimal short scan convolution reconstruction for fan beam CT. *Medical Physics*, 9(2):254–257, 1982. _eprint: https://aapm.onlinelibrary.wiley.com/doi/pdf/10.1118/1.595078.

[25] Todd E. Peterson and Lars R. Furenlid. SPECT detectors: the Anger Camera and beyond. *Phys Med Biol*, 56(17):R145–R182, September 2011.

[26] J. Qi and R. M. Leahy. Iterative reconstruction techniques in emission computed tomography. *Physics in Medicine & Biology*, 51(15):R541, 2006.

[27] Stefan Schaller, Thomas Flohr, Klaus Klingenbeck, Jens Krause, Theobald Fuchs, and Willi A Kalender. Spiral interpolation algorithm for multislice spiral ct. i. theory. *IEEE Transactions on Medical Imaging*, 19(9):822–834, 2000.

[28] Will Schroeder, Ken Martin, and Bill Lorensen. *The Visualization Toolkit–An Object-Oriented Approach to 3D Graphics*. Kitware, Inc., Fourth edition, 2006.

[29] Robert L. Siddon. Fast calculation of the exact radiological path for a three-dimensional ct array. *Medical Physics*, 12(2):252–255, 1985.

[30] S. Tong, A. M. Alessio, and P. E. Kinahan. Image reconstruction for PET/CT scanners: past achievements and future challenges. *Imaging in Medicine*, 2(5):529–545, October 2010.

[31] Henrik Turbell. "Cone-beam Reconstruction using Filtered Backprojection." PhD thesis, Linköpings University, SE-581 83 Linköping, Sweden, 2001.

[32] Heang K Tuy. An inversion formula for cone-beam reconstruction. *SIAM Journal on Applied Mathematics*, 43(3):546–552, 1983.

[33] P.E. Valk. *Positron Emission Tomography: Basic Sciences*. Springer, 2003.

[34] Y. Vardi, L. A. Shepp, and L. Kaufman. A statistical model for positron emission tomography: Rejoinder. *Journal of the American Statistical Association*, 80(389):34–37, March 1985.

[35] Ge Wang. X-ray micro-ct with a displaced detector array. *Medical Physics*, 29(7):1634–1636, 2002.

[36] Ge Wang, T.-H. Lin, Ping-chin Cheng, and Douglas M. Shinozaki. A general cone-beam reconstruction algorithm. *IEEE Transactions on Medical Imaging*, 12(3):486–496, 1993.

[37] Xiaohui Wang and Ruola Ning. A cone-beam reconstruction algorithm for circle-plus-arc data-acquisition geometry. *IEEE Transactions on Medical Imaging*, 18(9):815–824, 1999.

[38] Gengsheng Zeng. *Medical Image Reconstruction: A Conceptual Tutorial*. Springer-Verlag, Berlin Heidelberg, 01 2010.

Index